THE

INDIGO

PRESS

'The present-day ecological predicament entails the need to act immediately – within a few years – to avert the lasting wreck of our civilization and the climate that hosted it. No writer that I can think of is a better guide to this task than Richard Seymour: here is the quickest of thinkers, operating from the longest-term perspective. *The Disenchanted Earth* is an indispensable book.'

BENJAMIN KUNKEL, author of *Utopia or Bust*

'What thinker would you bring to an earth on fire? You would not want to leave Richard Seymour at home: he is essential company for an age of compound catastrophes. In these essays, he brings his trademark mix of psychoanalysis and Marxism, erudition and curiosity, pessimism and wonder, intimacy and sublimity to bear on the ecological crisis. They will keep you focused as the flames rise higher.'

ANDREAS MALM, Associate Professor of Human Ecology
at Lund University and author of *How to Blow Up a Pipeline*
and *White Skin, Black Fuel: On the Danger of Fossil Fascism*

'One of the most consistently brilliant and lyrical thinkers writing today turns his attention to the climate catastrophe – and the results are minatory and indispensable.'

CHINA MIÉVILLE, author of *October:
The Story of the Russian Revolution*

'Incisive. Truly radical. Full of short sentences and important insights. Seymour grabs environmentalism by the scruff of the neck. In thrall to its catastrophism, he also wants to save it from apocalyptic nihilism. You won't agree with all he says; nor should you. But you will always want to engage.'

FRED PEARCE, author of *A Trillion Trees:
How We Can Reforest Our World*

THE DISENCHANTED EARTH

Reflections on Ecosocialism and Barbarism

RICHARD SEYMOUR

THE
INDIGO
PRESS

THE INDIGO PRESS

50 Albemarle Street
London W1S 4BD
www.theindigopress.com

The Indigo Press Publishing Limited Reg. No. 10995574
Registered Office: Wellesley House, Duke of Wellington Avenue
Royal Arsenal, London SE18 6SS

A CIP catalogue record for this book is available from the British Library

ISBN: 978-1-911648-41-3
eBook ISBN: 978-1-911648-42-0

Essays in this collection previously appeared on Richard Seymour's Patreon

Art Direction by House of Thought
Cover design © Luke Bird

Author photo © Marta Corada

Cover photo © 1990 Amon Carter Museum of American Art
Photo © National Gallery of Australia, Canberra / Bridgeman Images

Typeset in Goudy Old Style by Tetragon, London
Printed and bound in Great Britain by TJ Books Limited, Padstow

CONTENTS

There are strange possibilities in every man. The present would be pregnant with all futures, if it had not already been informed with its history by the past. But alas, a one and only past can offer us no more than a one and only future, which it casts before us like an infinite bridge over space.

We can only be sure of never doing what we are incapable of understanding. To understand is to feel capable of doing. ASSUME AS MUCH HUMANITY AS POSSIBLE – let this be your motto.

ANDRÉ GIDE,
The Fruits of the Earth

translated by Dorothy Bussy

INTRODUCTION

A World Grown Old

11 JUNE 2021

'The twentieth century. Oh dear, the world has gotten so terribly, terribly old.'

TONY KUSHNER,
Angels in America

Don't be afraid, says Yeshua. Far more can be mended than you know.

FRANCIS SPUFFORD,
Unapologetic

I.

As in a disaster dream, catastrophe piles upon catastrophe. Just consider a few of the revelations from recent years.

A study published in the *Proceedings of the National Academy of Sciences* in 2017 warned of 'biological annihilation' as billions of populations of animals have been wiped out since 1900. In 2019, a survey published by *Biological Conservation* documented a collapse of the insect biomass at a rate of 2.5 per cent a year: a rate of extinction eight times faster than that of mammals, birds and reptiles. When they go, many animals will starve to death, with cascading effects throughout the food chain, while fewer plants will be pollinated and less fertile topsoil will be created. In 2020, a UN report compiled by 300 scientists warned that soil erosion endangers agriculture. Already, we have lost 135 billion tonnes of topsoil, which could take thousands of years to be renewed. In 2021, a study published in the environmental science journal *One Earth* found that bee species are going extinct, with 25 per cent of such species lost between 2006 and 2015, an urgent threat to pollination, and thus to the human taste for raspberries, apples, watermelon, cardamom, broccoli, apricot, coriander and pear.[1]

What these stories have in common is that none of them is spectacular in its effects – no wildfires in Arctic forests, no sudden disintegration and calving of giant ice shelves, no floods, no plague – but they all nonetheless describe processes that threaten the entire basis of human civilization. They bring unwelcome news of unacknowledged dependencies. Ecological sensibility has often been cultivated by pulling one's heartstrings about charismatic megafauna such as the polar bear or the North Atlantic right whale, and I'm rather fond of both myself. But without insects, those undesirable things we swat or shoo out of our kitchens, we would all be dead. It is not that we did not know of the immense historical significance of tiny creatures. Charles Darwin, towards the end of his life, wrote a little-known book on worms and their vital role in burrowing through and eating soil, thus providing the basis for vegetation, called *The Formation of Vegetable Mould Through the Action of Worms With Observations on Their Habits* (1881). He wrote: 'It may be doubted whether there are many other animals which have played so important a part in the history of the world, as have these lowly organized creatures.' We have known all that for some time. It is just that, for as long as those hockey-stick charts documenting rising water use, faster transit, more fishing and urbanization looked like a spectacular success story, we preferred not to think of it.

Who, you might be asking, is this 'we'? The pseudo-inclusive 'we' is one of the most obnoxious habits of the middlebrow male writer. To speak of 'we' in this context is to elide vast gaps in our relative ability to act. For example, the fossil fuel giant Exxon spent decades privately acknowledging the growing science of global heating even while publicly feeding denialism. Their ability to act, and their choice to engage in denial,[2] disempowered billions of human beings who didn't sit on Exxon's board, many of whom were fighting to stop the suicidal juggernaut. Research has established that just a hundred companies are responsible for 71 per cent of the world's carbon emissions,[3] a process in which the vast majority of humanity has little say.

Even to speak in general terms of human civilization being endangered is to gloss over the enormous difference between ecological devastation wrought in previous epochs, such as the mass killing of animals inflicted by the Roman empire, and the rolling cataclysm that hasn't stopped gathering pace since the Industrial Revolution. The issue is capitalist civilization. Capitalism, as the environmental historian Jason W. Moore puts it, is a 'multispecies affair'.[4] It achieves enormous amounts of throughput, securing a historically unprecedented amount of profit, thanks as much to the free work of pollinators and topsoil creators as to the human workforce. It depends on appropriating this as a 'gift', 'cheap nature', and on externalizing the costs of environmental destruction.

And yet, insofar as there is hardly anyone left on the planet who does not work for capitalism, does not buy commodities and does not depend on elaborate global supply chains for basic goods, 'we' are all implicated. Capitalism is something we all do, even if in very different ways, even if only as workers or consumers. To suppress fossil fuel extraction, to end the disastrous practices of agribusiness, to drastically curtail aviation, to stop the emissions and deforestation caused by the mass consumption of livestock and to establish truly sustainable fishing would require a drastic overhaul in the conditions in which billions live. No lasting change can happen without the buy-in of those billions. One would think, given the scale of the challenge, that there should be emergency public meetings in every village, town and city every week to thrash out solutions. Instead, given a pervasive sense of powerlessness and futility, the most common response is what psychoanalysts call 'disavowal': I know perfectly well that things can't go on this way but, because life is hard enough and I have bills to pay, I behave as if I don't. This is the emotional substratum for what Renée Lertzman calls 'environmental melancholia',[5] an undercurrent of sadness and thwarted mourning which can register in outward form as a defensive indifference.

II.

I know whereof I speak. These essays are a chronicle of my ecological awakening. As a young activist, I had little time for earth talk. Expressions of concern about the welfare of non-human animals, let alone climatic systems, generally aroused a defensive shrug. I acknowledged climate change but, while the 'web of life' is the irreplaceable foundation for all human endeavour, I tended to treat ecology as a subsidiary concern, one for the sorts of young activists who actively chose to be badly dressed (I merely defaulted to being badly dressed). What was the fate of the whales compared with stopping war or ending capitalism? I had even less interest in the natural sciences. Left-wing thought tended to be sociocentric, with chemistry, palaeontology, evolution, oceanography and zoology appearing, if at all, as interesting peripheries in the bibliophile's glut of history, political economy and philosophy. I was thoroughly, cheerfully, moronically insulated from the sense of danger.

Awakening came in the form of a grief catching up with me. The scene wasn't spectacular. Just a particularly warm, clammy winter, one of the warmest on record: there have been a lot of those lately. On Christmas Day, the fields and meadows of Trent Park were sprinkled with light, tepid drizzle rather than frost or snow. And, for some reason, this tiny vista gave me a glimpse of something, an awful sense of loss, that I could not look away from. In the recent annals of phenological chaos – spring coming ever earlier, polar perturbation plunging North America and Europe into deep chill in April, winter temperatures ranging from 20 to 35 degrees

Celsius in US cities – a warmish winter was scarcely a blip. Far less did it imply the potential collapse of the food chain, the flooding of coastal cities by polar melt, unmanageable wildfires, or ocean acidification threatening coral reefs – those underwater metropolises which are more productive than forests, savannahs, coasts or open ocean. But for some reason, all this knowledge which had been pushed into the depths forced its way to the surface.

This was not just an intellectual transformation, but a reformation of sensibility. It inflamed a passionate amateurish curiosity about matters on which I was utterly, stupidly ignorant: animal evolution, biogenesis, geology, marine science, animal minds, palaeoceanography, palaeontology. I am scarcely less ignorant today, but ignorance is no longer forbidding, and finding out is a labour of love, the word 'amateur' deriving from the Latin for 'lover'. And all this was not just a means to understand the dilemma of human beings in great peril, but a way to participate in the existence of things, as John Keats once said. I was looking, in other words, for a planetary sensibility. An experiential frame of reference encompassing the biosphere, a space that the Russian geochemist Vladimir Vernadsky defined in *The Biosphere* as a ring, perhaps measuring anything up to eighty-three kilometres deep according to modern estimates, spanning the ocean floor and the upper atmosphere. A space that is over 17,000 times older than the human race, whose global dominance is, like the ancient emergence of microorganisms in the ocean depths, a fluke. Another word for this type of experience is transcendence.

It often seems to be this way. We spend years becoming acclimatized to pollution, and desensitized to the loss of biodiversity, until a warp in the seasonal cycle, like the European heatwaves in the summer of 2019, assumes a prophetic significance. There is something about the palpable experience of climate change that is unsettling. To pun on Sigmund Freud's concept of the 'unheimlich' (uncanny), we could call it the 'untime-lich': the supernatural sense of being out-of-time. Even the Anthropocene is an attempt to name this experience, as it applies to geological deep time: the order of periods, eras and epochs accelerating as capitalist civilization leaves its mark in geological sediments, ushers in mass extinction for the first time since the dinosaurs, and threatens to end the cycle of ice ages which created the conditions for human thriving.

III.

Awakened in this way, it is difficult not to become a catastrophist. The Intergovernmental Panel on Climate Change (IPCC), for example, has consistently underestimated the rate of change in emissions, temperature increase, Arctic meltdown, sea level rise, ice sheet thaw, ocean acidification and tundra thaw.[6] In portraying

climate change as a smooth, linear process of temperature increases, it has neglected the variable of 'tipping
points' – drastic and irreversible climate transitions, with
cascading effects across the planet. An example would
be Amazon deforestation reaching such an extent that
the region was subject to regular droughts, or if there
was a large-scale die-off of coral reefs leading to the
collapse of marine life. When the concept of 'tipping
points' was introduced a couple of decades ago, it was
assumed that we were unlikely to reach such points,
since they would only kick in if the planet heated to
five degrees above pre-industrial temperatures.[7] That
estimate has since been progressively revised, and in
2019 an important study for *Nature* concluded that
many 'tipping point' thresholds may already have been
crossed.[8]

Equally, efforts at 'climate governance' have been
a marked failure. As David Wallace-Wells reports in
The Uninhabitable Earth, the vast majority of carbon
emissions in the entire history of humanity have been
released in the period since the Earth Summit in Rio in
1992.[9] The 1997 Kyoto Protocol, supported by the major
fossil fuel giants and signed by eighty-four governments
between 1998 and 1999, had almost no impact on
restraining emissions growth. Even the Paris Agreement
signed on 12 December 2015, which ostensibly committed signatories to keeping global warming at or
below two degrees above pre-industrial temperatures
(long thought to be the threshold of disaster), was

inadequate. The measures that governments agreed to would lock in a carbon-intensive energy system for decades,[10] allow fossil fuel giants to continue profitably to extract deadly fuels, and lead global temperatures to rise as much as a disastrous 3.7 degrees above pre-industrial temperatures.[11] Nor was there any enforcement mechanism to ensure that even these inadequate goals were met. The ink scarcely dry on that agreement, a long-brewing nationalist reflux propelled a series of leaders from the denihilist right, above all Donald Trump and Jair Bolsonaro, to power, thus accelerating emissions and deforestation even further.

The entire energetic infrastructure of modern civilization needed an overhaul. But because the fossil fuel corporations were so enormous, centralized, strategically key to modern economies and politically powerful, they could obstruct serious efforts to stop the coming disaster. And because governments were committed to an economic model, neoliberalism, that excluded the kinds of state-led industrial reformation that would be needed to suppress fossil fuels, permitting only ineffectual market mechanisms like emissions trading schemes, they were not well disposed to challenging the fossil fuel giants. The choices were thus between a liberal climate governance model that locked in deadly emissions, and a muscular nationalism of the right that rejected any limitations on fossil fuels as a Chinese conspiracy to cripple the economic development of the West.

IV.

Those of us who are of a catastrophizing bent are often told that we mustn't frighten people about climate change. Scare tactics don't work. Rather, we must tell 'stories' of change that inspire. Here, a few rote references to John F. Kennedy, Martin Luther King, Mahatma Gandhi and other worthies long since sterilized for the middle-market imagination, suffice to convey the sort of elevating rhetoric that is called for.

This is both patronizing and wrong in its premise. People are already frightened by climate change, and they should be. Fear is not inherently illegitimate. Contemplating the worst, such as runaway climate change triggering civilizational collapse, is not inherently stupid. If anything, we need more of George Orwell's 'power of facing unpleasant facts'.[12] Complacency and vague optimism, by contrast, are almost always stupid. At worst, they reek of unacknowledged terror. The problem is that most of us feel powerless to do anything about it because, on a day-to-day basis, we often are. And that powerlessness is aggravated, rather than assuaged, by the tactic of guilting people about their personal consumption habits.

Mainstream environmentalist organizations and news media love to give people listicles illustrating 'what you can do to stop climate change': don't waste water, fly less often, eat more vegetables and less meat. These are sensible measures which, if taken up by the wealthiest consumers, could make some small impact. However, the problem is one of aggregation. The logic of moral pressure that aims to change individual consumer decisions is that the market will aggregate the change in the form of price signals: as people fly less, for example, the slackening of demand should in principle lead to prices falling, airlines becoming less profitable to run and reduced economic pressure to build new runways. Yet airline tickets are already intensely subsidized by governments, who encourage and expand air travel and invest heavily in aviation infrastructure as a means to generate growth. As a result, domestic flights are generally much cheaper than the equivalent journey by train. The aviation industry is so strategically central to how states support growth, and how most large corporations do business across borders, that it has been protected from emissions reduction schemes.[13] And even when demand for air travel has slumped, as during the Covid-19 pandemic, governments have generally offered bailouts with no strings attached.[14] The same applies to agribusiness. Meat and dairy industries are already given whopping subsidies to keep prices low for consumers.[15] Consumer power is negligible, because consumer decisions are conditioned by what is available and incentivized. Unfortunately, alternative and more politically efficacious ways of aggregating individual will have been in crisis for decades. Civic society organizations, political parties, trade unions and other collective associations have been dwindling in both number and membership globally.

And yet it moves. In recent years, the salience of climate change has soared, with climate strikes led by school pupils and mass protests. In 2019, climate protests mobilized six million participants. Extinction Rebellion (XR), whatever the limits of its mobilizing strategy, brought civil disobedience to the heart of this movement. The Green New Deal, which had been on the agenda of a handful of economists, politicians and environmentalists when the idea was first mooted in the latter half of the 2000s, assumed global prominence as it became linked to electoral campaigns like that of Bernie Sanders. And as dangerous as the nationalist revival has been to human survival, its success has in part derived from the collapse of neoliberalism. The fact that Joe Biden, a very traditionalist member of the Washington establishment, has committed himself to significant green infrastructure programmes demonstrates that, given the right conditions, protest and disruption work. There is power in collective action.

In no way, of course, has the US political establishment broken ranks with the energy giants, nor is it even close to a serious reckoning with other sources of ecological destruction such as agribusiness and fishing, let alone a breach with perpetual growth. Moreover, the disposition of the People's Republic of China now represents a much more serious challenge than it once did. Notwithstanding its own rhetoric of 'ecological civilization',[16] it has invested so heavily in fossil fuels that its annual emissions are currently larger than those of all other developed countries combined. Its Paris commitment to achieve 'peak' emissions by 2030 gives it unlimited latitude for further expansion until that date, a clearly untenable position. Its most recent Five-Year Plan, covering 2021–25, made no commitment to reducing coal-fired power stations. Even where it has made such promises, it keeps right on building them, from Vietnam to Pakistan. And because it is a growing world power, it is building them across the globe and not just in China. Its investment in renewables and the extraction of rare metals necessary for solar panels is less an agenda for 'ecological civilization' than part of the same drive to industrial self-sufficiency and dominance: by the time it is ready to abandon fossil fuels, it will enjoy a substantial lead in the renewables-based economy.[17]

In short, the kind of global binding agreement that is necessary to stop the annihilation of life will require more intensive and internationally federated forms of protest and disruption than we have hitherto seen. As things stand, we remain on a course towards a 'hothouse earth', inhospitable for human civilization. And there are political tendencies afoot which would accelerate that course. Pre-eminently, there is the revival of a form of inchoate fascism whose response to ecological crisis is either hypertrophic denialism, accompanied by a drive to make the costs of ruin fall on the weakest in a social Darwinist pattern, or a green fascism which conflates conservation with a war against mutant or inferior biological forms. There is every chance, given the privations and difficult choices that ecological survival would mandate, that this form of politics will become potently relevant for millions of people denied real power, or a persuasive alternative to ecological austerity. There is every chance that we have already missed the opportunity to prevent some of the major disasters that will catalyse this suicidal political drive. There is every chance that we will miss the opportunity to arrest a destructive spiral that would be ruinous for human life.

However, the successes of recent years mean we need not be in thrall to our doom. And this is the real problem with ecological catastrophism: in stalemated situations, it can offer a dubious kind of satisfaction and even consolation. Think, for example, of the fatalist elements in the trend of 'deep ecology'. To imagine environmental apocalypse as though it were a disaster dream in which we are fully present to witness the terrible end would be rather like the ubiquitous fantasy of attending one's own funeral. If you're there to see it, you're not really dead. Unconsciously, Freud argued, no one believes in their own mortality. In the long run, as John Maynard Keynes argued, we're all dead.

LETTING GO

15 APRIL 2017

Why add more words? To whisper for that which has been lost. Not out of nostalgia, but because it is on the site of loss that hopes are born.

JOHN BERGER,
And Our Faces, My Heart, Brief as Photos

This is the Hour of Lead –
Remembered, if outlived,
As Freezing persons, recollect the Snow –
First – Chill – then Stupor – then the letting go –

EMILY DICKINSON,
'After a great pain, a formal feeling comes'

Robert Macfarlane's *Landmarks* is a 'counter-desecration phrasebook': a vocabulary for valuing what we have just as we are about to lose it, just as we are losing it, just as we have already lost it.

It is as if the living world, of shivelight and suthering tides, of desire lines and whale's ways, of glaise and drindle, sumping sea lochs and high headlands, could be saved through redescription. As if it weren't already too late.

The last fourteen months have, one after another, broken global temperature records. Floods and droughts begin to assume biblical proportions. Thousands of species disappear, forever, each year. Even in the mildest prognostications, they will disappear faster and faster.

If temperatures increase more than 1.5 to 2.5 degrees above pre-industrial levels, the IPCC estimates that 20–30 per cent of species risk extinction. If the temperature increase exceeds 3.5 degrees, the range is 40–70 per cent. We are already at 1.3 degrees, and 4 degrees is approximately the current projected temperature increase by 2050, even if the Paris Agreement survives.

As the rate of acceleration increases, so does the probability of chaos. Scientists use the metaphor of 'uncharted territory' to describe this, since all we know for sure is what we are losing. What will never, ever be seen again.

Walking, in this way, becomes an urgent voyage, a pilgrimage, a visit to a dying patient. A stolen glimpse of what might have been won, had the earth ever been a common treasury.

But as the psychoanalyst Christopher Bollas points out in *The Evocative Object World*, what we find in the environment is our own unconscious life – not in its narrative, or in its scenery, but in keywords, objects. The more abstract, nonsensical and formless the terrain, the more we can project into it, and the more evocative it seems. Nothing is more suggestive than what theologians, following Psalm 22, call 'the night season'.

What you find in the burnt edge of a cool morning, the summer shimmer of riparian wetlands, clouds the size of cities soaking in a blue pool, or even in the literary outdoors – the cold mountains of Hanshan, the freezing Yukon of *The Call of the Wild* – is unconscious meaning.

Worlds of independence, adventure, possibility, decivilization, worlds teeming with potential, closer to birth than death. Oceanic immersion, the feeling of being held, protection. Phobias and anxieties. Screen memories. These private meanings always open out into public meaning. What Lertzman calls 'environmental melancholia' begins with lost worlds. Melancholia is a kind of freeze. Mourning is movement, and if you can't mourn, you gather frost.

One of the biggest obstacles to mourning is that we can't face our ambivalence: the extent to which we hated the lost object of our love. The ambivalence is complicated. On the one hand, it seems, no matter how much they meant to us, we're always in some part of us glad to be shot of them. On the other hand, we also hate them for no longer being there. And there are the unconscionable pleasures and benefits that accrue from their absence.

We can hardly help being ambivalent about what we call 'nature' and its nemesis, fossil capital. The former means desperate, hard, labouring lives and early deaths. The latter, to the extent that it is coextensive with industrialization, promises comfort, central heating, celerity.

So what is the greenhouse defrosting of Arctic sea ice, the bleached death of a coral reef and the disappearance of thousands of species every year compared to air travel, moon voyages, genetic science laboratories and the internet? What is the silence of the remote croft, or the murmur of the forest, compared to rising life expectancy and falling infant mortality?

The other side of this ambivalence, the nocturnal side, is the knowledge – because this is no mystery, and anyone who wants to know already knows – that we are preparing a mass wake for the human species. It is a planned obsolescence. There are some hubristic billionaires who, by investing in survivalist Xanadus, fancy they will survive the collapse of the food chain and the destruction of habitable territory. Few have the luxury of that conviction. So, put another way, the questions above become: what is species death, compared to another fifty years of life for capitalism?

It is useless to berate the insufficiently woke. We are all sleepwalking, and all half dreaming, even if we dream of being awake. We are all hastening towards the last syllable of recorded time. And the point of melancholic subjectivity is that we are already berating ourselves. Our experience of powerlessness in the face of loss, and of isolation before gigantic, tectonic forces, has already become our mantra of self-hate. Adding reproach in the name of the future would only accentuate our resentment of future generations, and our desire to punish them.

But if mourning is movement, it is also work. The work of mourning is not the same thing as the sharp icicle stab of grief you might feel, while walking, when you suddenly realize that some day and soon, nothing that looks like this world will exist. It is the painful, laborious task of revisiting each memory, each thought, each impression, of what has been lost and, like Edgar Allan Poe's raven, meeting it with the judgement 'nevermore'. Mourning is not an uplifting process. It is a kind of despair, because it means giving up. First chill. Then stupor. Then the letting go.

Only when we can separate the object that has been lost from what has been lost in it, do we recover. In other words, we give up without giving up. We fully and relentlessly recognize the loss, but we hold on to the qualities we saw in the lost object, because we think we can find a way to revive them in a new passion, a new attachment. We despair, but we do not submit.

'Despair without fear, without resignation, without a sense of defeat,' Berger called it, speaking of the Palestinians and their al-Nakba. 'Undefeated despair.'[1]

DROPPING LIKE BEES

15 AUGUST 2017

I'm in Iceland, watching bees feverishly court the foxgloves in the stinging cold.

They're working as though there weren't a billion tiny specks of rain drifting down on them like smoke. As though it really were summer.

And I am abruptly reminded of the need for this frantic labour. If the bees were to disappear as a species, humans would join them. We would drop like bees, by the billions, famished.

We depend, albeit obliviously for most of our existence, on the bees for a huge amount of the food we eat. The mere work of pollination is worth billions – of people, and pounds. It is worth tonnes and tonnes of exportable crops.

The sudden, sharp collapse of bee colonies across Europe and North America over the last century has been, gradually and reluctantly, correlated with capitalogenic climate change. A body of research tracking the relationship has begun to develop. Certain pesticides may also accelerate the problem, beyond the point of repair.[1]

The interesting thing about the phenomenon of colony collapse is that it resembles an abrupt and irreversible work stoppage. The worker bees simply quit, walk off the job, leaving enough food for the short-term survival of the queen and infants.

We all sort of knew of, but took for granted, the sexual and reproductive labour of pollination. Until the possibility of its sudden withdrawal brutally forced us to face up to an unacknowledged dependence. One species death brings another in its wake.

II.

Let's say it again: we sort of knew. And we sort of know about multitudes of other ecological dependencies, even if we proceed as though we didn't know.

The word for sort-of-knowing but ignoring is disavowal. In psychoanalytic terms, we disavow in order not to admit our castration, our dependence. And this particular disavowal is an operation of capitalist social relations.

It is not that it would be a good idea to re-enchant the earth, even were that possible. But disenchantment, as Theodor Adorno and Max Horkheimer have shown from one perspective, and Carolyn Merchant from another, was part of a gigantic civilizational rupture as the sixteenth century turned into the seventeenth century, bringing new modes of oppression and exploitation with it.[2]

The augmentation of the early modern state as it struggled to manage the emergent capitalist system. The acceleration of the Reformation into a continent-wide war that consumed eight million lives, produced a demographic crisis and triggered the formation of a new states system. The enclosures and witch-hunts, the re-regimentation of gender on the basis of a division between public and private. The transformation of the animisms, magical practices and alchemies of the Renaissance into the mechanistic, experimental sciences of the Enlightenment.

The gains of this continental cataclysm, of course, need little elaboration here. We enjoy perpetually longer lives and expanding capacities, mobilities and literacies, and perhaps even the possibility of human emancipation before human annihilation, because of the progressive part of that explosion.

But, bringing with it a new set of social relations, it also brought with it a new set of conceptual distinctions and dichotomies. Above all, the creation of 'Nature' as a distinct and subordinate realm of being, over which 'Man' enjoyed dominion. And if Francis Bacon liked to imagine 'Nature' as a woman, to be interrogated, chastened and brought under control, the division being envisioned here would see women, workers and black and colonized subjects placed firmly in the camp of 'Nature'.

If the disenchanted earth, atomistic and mechanistic, was finally regarded as being so altogether available for domination, it was because it had been deprived of anything that could be regarded as agency. It was a raw material, potentially resistant, but otherwise strictly dependent and subordinate.

III.

In this way, capitalism obscures its own conditions of possibility, even as the screen image of capital as a sleek, immaterial, weightless spirit is perpetually obscuring its vulgar agrarian origins, its basis in the exploitation of plant, animal and human labour.

The disavowal of what we sort of always knew has consequences. If we cannot simply re-enchant the earth, we need to rediscover at the level of theory what has been blotted out of everyday perception.

This starts with the acknowledgement that, as Moore puts it in *Capitalism in the Web of Life*,[3] capitalism is a civilizational order that is 'co-produced by humans and the rest of nature'. It depends as much on the unpaid work and energies of forests, rivers and wind as on the unpaid work of women and slaves. Capitalism is a 'multi-species affair'. The bees work for capitalism.

There is, as Moore suggests, not only the 'socially necessary labour-time' of commodified labour, but also the 'socially necessary unpaid work' of uncommodified labour, which 'crosses the Cartesian boundary' between 'Human' and 'Nature'.

To make all this labour happen on terms commensurate with capitalist production, capitalists and states have to take hold of, observe, measure, classify and code all the various 'natures', human or not. They have to subject them to a capitalist grid of intelligibility, which is the grid of commodity production. All of these processes, wherein different forms of 'nature' are converted into preconditions for capital, Moore calls 'abstract social nature'.[4]

This – drawing directly on Donna Haraway's A Cyborg Manifesto, demystifying and dismantling the nature/culture opposition – assigns its findings a specific theoretical value within Marxism. And in so doing, it gives the term 'capitalocene' its proper conceptual basis, without which it would simply be a sarcastic rejoinder.

But it brings us to this. Capitalism's 'law of value' was always 'a law of Cheap Nature'.[5] And yet, of course, cheap nature was always a fiction. 'Abstract social nature' organizes its exploitation so that its costs are externalized, driven outside the circuit of production: but they are still costs borne somewhere. And we are beginning to see where: they have been piled up somewhere in the future, for generations unknown to encounter as their cataclysmic end.

From various directions, the strains are showing, and revealing themselves to be potentially terminal. The possibilities of extinction multiply. In our thousands, in our millions.

OPARIN'S OCEAN

24 OCTOBER 2017

The peaks of the Alps used to be a seabed.

Three hundred million years ago, the jagged rocky spires that are currently being ground down by wind and ice were under the sea of Tethys between the ancient continents of Gondwana and Laurasia.

They were pushed up out of the waters as the Eurasian and African tectonic plates crushed against one another. By now, the peaks have ceased to rise, and the same rocky surface which was once the floor of a teeming marine life is host to snow leopards, wolves, marmots and golden eagles.

As the geologist Charles Lyell observed, once you make room for time, all sorts of freakish effects become explicable.[1] The Grand Canyon looks like nothing that could be achieved by mere erosion. But once you assume that we live on an old planet, you can extrapolate gigantic transformations from tiny changes barely perceptible to a human eye. It was on this foundation that Darwin was able to formulate a whole theory of life.

With time, you didn't need an intelligent designer to explain variation among living creatures. The compulsion of natural selection would do the work. We now think that evolution can happen far faster than we once assumed it did. Bernard Kettlewell's experiments with peppered moths indicated that selection in industrial areas was working extremely rapidly to promote darker moths, as opposed to the 'peppered' moths whose look stood out against besooted tree bark.[2] The agent of selection was birds picking off the less disguised moths. Similar findings were recently published in *Scientific American* about the great tit, whose beaks have been getting longer over a relatively short time.[3]

Nonetheless, most of the work of evolution has been slow and brutal. For example, according to some palaeontologists, even the first animals didn't evolve until 570 million years ago, when soft-bodied Ediacaran fauna first appear in the fossil record. And this may be a clue to an ancient question: how did life begin?

How, to put it in other terms, did geochemistry become biochemistry? What made the difference between inorganic chemicals, reacting with one another more or less harmlessly on the planet's surface and making no apparent impact on the great rock faces, and the first primitive organism?

To even ask this question is to presuppose an answer to another: what is life? There are dozens of definitions offered by professionals and authoritative bodies, none of them the same. For the sake of a spurious simplicity, let us assume NASA's definition: 'Life is a self-sustained chemical system capable of undergoing Darwinian evolution.'

All life forms are chemical systems. They all grow and sustain themselves by gathering energy and atoms from their environment: metabolism. They all make copies of themselves before their cycle is finished: repro-duction. And they all adapt to their environment under selective pressures: variation. This is the minimum for anything we call life.

But how did this once black, hot, volatile, anoxic planet ever come to host anything like this? Anything like us? We are accustomed to thinking about this as though it had to be a sudden, qualitative rupture. And indeed, purely in logical terms, there is a qualitative difference between non-life and life. Darwin, in fact, had speculated in a letter to his colleague Joseph Dalton Hooker on a possible answer that was evolutionary: 'If (and oh, what a big if!) we could conceive in some warm little pond with all sorts of ammonia and phosphoric salts, light, heat, electricity, etc., present, that a protein compound was chemically formed, ready to undergo still more complex changes.'[4]

In other words, relatively simple molecules must have, under certain conditions, been combined and concentrated in a liquid, and exposed to some energy source, thus producing more complicated, prebiotic molecules, such as proteins, sugars and lipids.

However, Darwin seems to have abandoned this thesis, and it wasn't picked up again until 1924, when the Russian geochemist, Alexander Oparin, published his thesis *The Origin of Life*.[5] In it, Oparin argued that life must have been a product of chemical evolution within the oceans. The seas were the primordial, chemical 'soup of life' in which simpler, abundantly available chemicals could form more complex hydro-carbons, and thus the basis for organic matter. It would just require concentration, and some source of energy, such as lightning or UV radiation from the sun. This idea gained the support of the British scientist J.B.S. Haldane.

When the pamphlet became a book in 1936, Oparin garnished the thesis with the language of 'dialectical materialism', as against Western imperialist scientific idealism. Making peace with the emerging dictatorship, even to the point of supporting Lysenkoism, enabled Oparin to build a huge institutional structure in which to pursue his work. But it also ultimately ruined his career, when the pseudoscience of Lysenkoism was repudiated after Joseph Stalin's death. Nonetheless, Oparin's thesis is a materialist thesis, and a logical development of Darwin's idea. And it's an irony that its partial confirm-ation took place in the same year that he began to lose clout in the USSR.

The famous Miller–Urey experiment was chiefly the work of an ambitious young chemist called Stanley Miller. Keen to make his mark, he came up with an elegant experimental way of testing Oparin's thesis. This involved two glass bulbs, a few inches in diameter, connected to one another by tubes. In one bulb, there was water heated by a flame held underneath. In the other, a mixture of gases exposed to electric sparks from two electrodes. Within days, the water changed colour, and eventually there accumulated an organic black slime on the glass walls. The slime was a mixture of amino acids, lipids and carbohydrates.

This experiment has been successfully repeated, over and over. It has often produced highly complex RNA and DNA molecules. So it seems plausible that it describes something like the process through which life's building blocks were created. And, indeed, once this is established, the next step can be taken.

A protolife might have begun when these more complex molecules began to make copies of themselves. There are various models for replication, such as the reverse citric acid cycle, or self-replicating RNA. However, once a molecule is capable of reacting with others to produce a duplicate of itself, it can begin to populate the oceans quite quickly. Assuming that the molecule duplicates itself once a day, the first month might see about a billion such molecules formed. This would literally be a drop in the ocean, since a single drop of seawater contains about ten million viruses. But within a year, you might have 7.5 billion googol molecules, something closer to the scale of an ocean. At a certain stage of concentration, it is plausible that selection might produce a mutation capable of interacting with and exploiting the mineral environment, thus developing greater and greater complexity, until cells emerge.

There are, of course, problems with this picture. The conditions in the glass bulbs were far from those on earth, with its complex and overlapping climatic, oceanic and other planetary cycles. The geochemical environment of early earth was far more complicated than the experiment allowed.

The experiment also presumed a degree of chemical concentration that would not normally be present on the planet. However, there are ways in which planetary processes, such as freezing, could have produced the necessary concentrations.

There is also the problem that protolife does not seem to have left any fossil residue at all. Perhaps, as soon as the first cellular organisms existed, they consumed their evolutionary predecessors. But this recalls the rule about the evidence of absence.

There is also a rival hypothesis, one which was prompted by the discovery of deep underwater microbial communities, in the darkest regions of the ocean. There, it is thought, hot underwater volcanic crust and oceanic geysers, interacting with mineral-rich water, might be enough to produce the materials for life.

It could be both these and many other things. Life could have more than one origin. It is even probable that there were several different prebiotic processes, which could have but did not produce life. Notably, both of the scenarios mentioned above involve chemical evolution within a prebiotic soup over hundreds of millions of years, fuelled by energy (from the sun, or volcanic heat, or lightning).

Oparin's ocean has, among its major virtues, this characteristic: it grounds no teleology. It cannot be anything but a fluke, and one against all odds, that life actually emerged from the primordial stew. The conditions required to make it happen certainly must have existed at some point, but as any palaeontologist or geologist could tell you, they can in no way be inferred from the original state of the planet in the so-called Hadean Eon.

And of course, the corollary holds. The extent to which life holds on, to which the planet doesn't simply change dramatically and snuff it all out forever, is the purest fluke going.

UNWORLDLINESS

24 NOVEMBER 2017

I.

Worldliness is death. Disenchantment is mystification.

In the very morning of life, animals began to flourish and multiply on the slopes of underwater volcanoes, thriving on the choking black smoke, at temperatures and in chemical environments that would kill most creatures.

In life's midnight, when the musk of the last rose is eternally blown, and with it the last utterance of a human wish, there will be animals who will still survive in baths of boiling acid. Unlovely, uncharismatic, but far better adapted than the species they have left behind in deep geological time.

And if they are wiped out, there may be, who knows what, who knows how, another geochemical reaction somewhere, somewhen, that leads to new biochemistry. New metabolism, new mitosis, new meiosis: new life. Perhaps, given the logic of convergent adaption, new hominids.

With any luck, these new life forms will discover the trace of your life, not as fossilized bones or an ultra-thin layer of geological sediment, but in energy-rich fossils that they will discharge into the heavens in a billion fine particles. The revenge of deep history.

II.

Thus saith the Lord God; Come from the four
winds, O breath, and breathe upon these slain,
that they may live.

EZEKIEL 37:9,
on the Valley of Dry Bones

Friedrich Nietzsche, amid one of his characteristically
bracing attacks on Christianity in *Ecce Homo*, said
that faith was 'world-calumny'.[1] If God 'so loved the
world' that he sent his son to be sacrificed for its sin,
Christianity hates the world, because the mutilation of
Christ is its handiwork.

The 'world', in this sense, has more than one sense.
It is the planet, of course, ravaged by hell. It is the flesh,
blighted with the weakness of original sin. And it is the
social world, the historically produced world of human
beings, with its attainments, pleasures and productions.
When you are thoroughly sick, heart-sick of the world,
you may indeed hate all three.

To despise the world in any sense, however, must of
necessity be to love another world that exists, as yet,
only in the navels of our dreams. A world that has not
yet been conceived, let alone born. World hatred is
always implicitly utopian, presupposing the existence
of another world where things work better, even if there
is no way to say what it would look like. Even if it is
unspeakable.

This is where the dimension of faith enters politics, and where a certain kind of worldliness has to be abandoned. There is always a point beyond which our claims, however materialist our premises, can have no rational foundation. We cannot prove that this other world is possible, and beyond some preliminaries we cannot even speak sensibly about what it would look like – Marx was sceptical of utopian blueprints for a good reason. Faith connects us to the unknown, to that part of existence which cannot even be said.

Worldliness is a kind of narrow-mindedness. By claiming that this world is all there is and ever can be, and that its pursuits and benchmarks are therefore all that matter, we feign omniscience – 'that will never happen' – and thus shut down whatever is unspeakable and unknowable. We impoverish our repertoire, becoming – as Daniel Bensaïd said of post-Cold War historians – 'notaries of the accomplished fact'.[2]

And in so doing, and in a way, we actually cease to engage with the world, since we no longer expect it to surprise us. It becomes dead, inert, and so do we. As dead as dry bones.

III.

This world, which one sometimes calls civilization, only half exists.

Civilization, Freud wrote in *The Future of an Illusion*, is a means of waging a war against nature. But the feeling of mastery that civilization gives us is illusory, and subject to 'the elements which seem to mock at all human control' and which rise 'against us, majestic, cruel and inexorable',[3] the return of the repressed.

If mastery is a defence against the anxiety of dependency (a dependency acknowledged in such maudlin locutions as 'Mother Nature' and 'Earth Mother'), it can only give us new anxieties. The control humans have achieved 'over the forces of nature', says Freud in *Civilization and its Discontents*,[4] is such that 'they would have no difficulty in exterminating one another to the last man'. At present, in this, the sixth mass extinction, it is the other species we are exterminating. The scenic route to our own extermination.

The myths of the age, of the sovereign free individual, the entrepreneur as rugged pioneer, are but death hymns. We may need new myths which recognize dependency, including its dark side. The problem with the Earth Mother myth, in this respect, is not necessarily its mawkishness. It is that, like all idealizations, it suppresses the ambivalence that we always have about mothers, because they are always ambivalent about us. The same Earth Mother that bears life also periodically chokes most of it to death. Shove the Earth Mother in people's faces too long, and they'll be yearning for matricide.

The earth that we know, on which the life that previously thickened and bloomed in seas and reefs made landfall amid bountiful vegetation, is barely 300 million years old. Yet this same age, the Carboniferous period, which gave us a great terrestrial flourishing and filled the air with oxygen, bequeathed us a strange, poisonous hyperobject, fossil fuel. A fossil to fuel our fantasies of control, and then ruthlessly reduce them to mockery. Donald Winnicott's 'good enough mother' came loaded, heavy, with Michael Eigen's 'toxic nourishment'.[5] We need a better myth than this.

IV.

What we call disenchantment is closely related to omniscience and ideas of mastery.

In a disenchanted world, everything is in principle calculable, intelligible in the light of scientific and rational principles. If we haven't yet worked out the explanation for something, there is no doubt that we could. All surprises are pre-emptively accounted for.

But if the precondition for disenchantment is the Baconian idea of science as the human enslavement and plunder of 'nature', where 'nature' is separate from human beings, then it is a mystification. Ultimately, as Moore has written, disenchantment constitutes various 'natures' through the screen of capital.[6] Whatever does not produce value for capital is passive, 'cheap', and there for plunder.

The philosopher Jane Bennett argues in *The Enchantment of Modern Life* that we would benefit from being less impressed by the story of disenchantment. It is a tale that modernity has created about itself, its origins and its telos, but it is both analytically and ethically suspect. Analytically, because it treats matter as merely inert, in order that it be masterable. Ethically, because it forecloses the kinds of experience that would be most likely to foster generosity.

Following Nietzsche, Bennett contends that pity, guilt, victimhood and resentment are poor motives for unlearning harm and acquiring ethical generosity. Most likely, these affects make us more narrow-minded, more defensive, more likely to want to control and police the world, and to find our satisfactions in the miseries, small or large, that we inflict on others. The same may be true of the environmental melancholia of which Lertzman has written.

The capacity for enchantment, hard-won in a bitter world, is a more plausible alibi of justice. To be enchanted by something in the world is to be simultaneously surprised, charmed and unsettled by it. Something in the everyday suddenly impresses us as novel and uncanny. It absorbs us into its uncanny particularity for a moment, suspending the usual temporal flow of capitalist work-time, putting our worldliness into question.

v.

The question tacitly posed by the experience of environmental melancholia is: what if it's too late?

There was one moment in the last century when the apparatus of extinction was seriously threatened. When it was the capitalist mode of production whose survival was at stake. And what if the consequence of capitalism's survival for another hundred years means lights out for the species, or for earthly life itself?

What if, in fact, all that we are doing now is making arrangements for our ultimate interment as geological residue? What if this unplaceable sadness sticks to the species, dimming the diurnal light, crowding out the stars, interceding between life and the living so that it is as if we're already half gone, so that the things of the world pass before us as if from another lost dimension, in a funereal procession? Like a tyrant, this melancholy demands constant satisfaction, reminding us never to be too exuberant, never to be too kind, never to be too free, never surprised or moved by life. Half gone, in a shallow grave, life's traffic thundering overhead: altogether too much. Is it worth this much noise, if we're only a lot of dry bones? These are the sad passions of the Anthropocene.

Then melt the tundra frost, turn the oceans acid, let the methane rip and make the Arctic boil. Let billionaires be our gods, with offshored claims on our disappearing future. Let oiled vulgarians have eternal boyhood, and those who can live like malignant kings. Let a whole race of Trumps emerge to be entranced with its own ordure. Until a runaway greenhouse turns Earth into Venus. Until the last amber-honey flush of a stray autumn breeze becomes solar wind and fire.

There is, besides, a certain covert satisfaction, a compensation, to be had in contemplating all-out destruction. It is like imagining your own funeral. As Freud pointed out, unconsciously, none us really believes in our own mortality. We don't think we can die. Fantasy is structured around an impossibility, in this case being the witness to our own absence. So, in relentlessly and pitilessly confronting eco-death, we find that the secret consolation is a reinstated immortality and omnipotence. Knowledge alone does not save us.

VI.

If the danger of disenchantment is that it effectively leads us ever deeper into a logic of nihilistic plunder, and suicidal species-loathing and melancholia, there is no refuge in a glib enchantment.

There is always a point where ecocriticism becomes eroticism, yet the capacity for an enchanting encounter can always be usurped by merely another romantic cliché about 'nature' and the 'blue planet'. Enchantment risks becoming its opposite. By producing nature as a kind of reliable 'experience', an eco-Disney which promises you the same thrill each time, we reproduce the screen of capital, and the logic of disenchantment.

The BBC's extraordinarily popular David Attenborough-hosted *Blue Planet* programme, now in its second series, is the most sublime example of the tendency. In this, as in *Frozen Planet*, *Planet Earth* and innumerable others, Attenborough is a syrupy, benign, white, patriarchal god, whose voice booms warmly from the earth, commenting on the heart-wrenching beauty of his creation and its doings.

It is gorgeous, enthralling, and at times unsettling. It engages unconscious desires and dreads, solicits projections and identifications, creating the most inviting tableaus through which to engage one in dreamwork. At times, it is truly uncanny.

Blissful oceanic immersion swiftly becomes terrifying. Predators and unsettling briny bric-a-brac suddenly appear, spotlit in the dark depths. For a few seconds of footage, the cephalopod, a truly strange, evolved intelligence, makes a jet-propelled sweep of the abysmal ocean floor: and you encounter your awful, inscrutable double.[7] Galumphing on rocky coasts, elephant seals, hypertrophic, rasping, roaring and rending one another, spewing steam and spilling blood, crushing the young with their fat, alien-worm bodies, show us our own stupid drives.

As in pornography, identifications switch back and forth, between the polar bear and the leopard seal, the harlequin shrimp and the starfish, the orca and the sea otter, the Arctic fox and the guillemot chick, the hunter and hunted, the eater and the eaten, the fucker and the fucked. And throughout, the snuff footage, the kills, the carcasses, the carrion scavengers, betray the silent call of Thanatos lurking within this genre of softcore ecophilia.

Yet wherever the truly uncanny emerges, world-liness seeps back in through every crack. No matter the species, no matter the climate, from Arctic fresco to tropic mural, from peak to trench, arthropod to mammal, Attenborough's tutelary voice interrupts with an anthropomorphic analysis of its sexuality, aggression, rivalry, affection and kinship. There is home, there is family, there is property and theft, there are brutish males and difficult, impossible-to-please females, absent fathers, and cubs, chicks and calves gently bitten, beaked or borne up on megapterine backs by devoted mothers. Modern social attitudes, values and affective states are imputed to all manner of creatures.

This is tacit reassurance that, brutal as it may be, threatening and alluring at times, nothing in the world is really other, nothing points to a beyond, and everything that lives is just like us. There may be other beings, but there are no aliens. There may be other planets, but there are no other worlds. Only this, a deathtrap.

There is, thankfully, a limit to anthropomorphism, a limit to worldliness. Some creatures can't be human-ized. That strange hybrid, plant-animal, sexual-asexual, individual-community, the reef-building coral, has no 'personality' to speak of.

Thousands of polyps, sharing the same exoskeleton, which is excreted, and a single stomach cavity, somehow build the most efficient nutrient-sink in the ocean, oases of life in marine deserts. They are more productive than the forests or savannahs or coasts or open ocean, home to a quarter of marine life. And destinations for all sorts of creatures, from manta rays to sharks to sea turtles looking to be cleaned of their parasites and dead skin, which some reef animals eat.

These reefs, of course, are underworlds of plunder and prey, watery graves for millions of creatures, where the nightlife is only good for carnivorous predators. After each sunset, the terrors, the boiling feeding frenzy, the shower of scales and fine particles. But still, life goes on more abundantly than anywhere else in the ocean. The reef flourishes through dependency and cooperation, in a completely alien way.

VII.

Or speak to the earth, and it shall teach thee:
and the fishes of the sea shall declare unto thee.

JOB 12:8

Enchantment has an etymological link with singing, poetry and prayer. As Dana Gioia writes, the rhythm and ritual of poetry aims to awaken, and give form and force to desire. To incant, to recite, is to wish.[8] If, as Walter Benjamin put it, dreaming has a share in history, enchantment has a share in the future.[9]

Unworldliness, or other-worldliness depending on your point of view, might be compared to what Keats described as 'negative capability'. To be enchanted, you have to be ready to experience something without immediately trying to make something of it. To speak of it is only to half speak. It is to merely musically gesture at a knowledge that is wholly half-knowledge.

It might also be a kind of faith. Something in the world unspeakably, impossibly, unsettlingly gives form to a desire, a future, that does not yet even have words, let alone a world to inhabit. To be in a place where one can apprehend such a thing, however, is to have made a secret wager that other worlds can exist. It is to make faith with a dimension of beyond that only gospel, song, poetry or prayer could speak of.

This is the 'sense of unsayable excess' that China Miéville has written about in the *Guardian*.[10] The 'blank white space' on the front cover of *Pravda* in those July Days adverts to something uncanny, a breach; words unsaid, strategic calculations stillborn, possibilities left dormant, lives unlived, worlds unborn, and all that could never be said in any words whatsoever, all clamouring and jostling for that little space.

What that little blank space said, it said apophatically: The ancient tyranny of the accomplished fact is upended. The machinery of extermination is unsettled. Nothing is known any more, calculations fail, mastery dies. Dust is flesh, and the slain live; the word of it shakes them in their shallow graves so that they, who slept through cannonade, fire and the death-drop of millions, awake and bloom. The earth is not your deathtrap; your earth is beautiful, and it is still the morning of life.

This was the news, unconsciously yet ecstatically transmitted, scorched across the land like the very flame of Yahweh, from east to west, from west to east, to farms, factories, frontiers, sailors, soldiers and soviets, on 4 July 1917.

THE ATOMIC GENIE

11 SEPTEMBER 2018

The question is not whether we're for or against nuclear power. The earth was forged by nuclear fusion, a miraculous by-product of the nuclear waste from an exploding star.

Keith Barnham, ecologist and physicist at Imperial College London, argues in his book *The Burning Answer* that we exist thanks to the nuclear energy from the sun, which drove the planet's mineral evolution from geochemistry to biochemistry.[1] It made plants that produced oxygen as a waste product, on which mammals subsist.

The question is whether, instead of building nuclear reactors on earth, we can rely on the nuclear fusion reactor at the core of the sun. This is not a socially neutral question. Recently, the *Economist* highlighted a fatal problem with solar energy. It's too plentiful: 'Because additional solar capacity will produce power at times when there is already a glut, returns to further investment in solar capacity will decline.'[2]

To be precise, solar energy is too plentiful to be profitable. Although intermittently, at peak times, there will potentially be a huge surplus of energy. The better the solar infrastructure, the worse it will be for returns on investment.

Luckily, our friend the atom comes to the rescue. The *Economist* is heartened to note that new legislation in the state of California allows for the possibility of nuclear energy being treated as a 'zero-carbon resource'. There are companies lining up to invest in nuclear energy projects. Admittedly, they rarely get very far without huge state subsidy, as with Hinkley Point C in Somerset. And states that are addicted to nuclear energy are likely to be those with nuclear weapons or an interest in procuring them. But it is obvious enough why the *Economist* would prefer crony state capitalism to unprofitable abundance.

II.

But it isn't just the *Economist*. Support for nuclear energy goes right to the heart of the climate change establishment. The IPCC has changed its position over the years. But its Fifth Assessment from 2014 was perhaps the most optimistic it has ever been.

Why, between 2011 and 2014, did the IPCC go from minimizing the role of nuclear energy to extolling it as a key element in climate mitigation? The short, surprising answer is that it adopted the framework of the International Atomic Energy Agency (IAEA), a body that exists to promote civil nuclear energy use.[3]

In its Fifth Assessment, the IPCC cites the IAEA frequently, while ignoring the abundant energy literature that is critical of nuclear power. The IAEA, naturally, describes nuclear energy as offering 'practically unlimited energy resources' for dealing with climate change. Its model for 'sustainability' is industry standards of 'best practice'. The IPCC's acceptance of this framework undermines its non-prescriptive, advisory role, and suggests that it adapted to the existing commitments of policymakers. Nonetheless, it has meant that industry assumptions about sustainability have been legitimized at the highest level.

The UK's Climate Change Committee (CCC) agrees that nuclear power must be central to decarbonization.[4] And the government is now committed to a major expansion of nuclear energy capacity over the next twenty years.[5] This commitment comes at considerable cost to the taxpayer. The UK government has a subsidy scheme for nuclear reactors, which will involve the public funding the difference between wholesale electricity prices and the high, fixed cost of nuclear electricity.[6] That scheme, which will benefit energy firms like EDF, was approved by the European Commission. So institutional support for nuclear energy expansion as a bridge to a post-fossil fuels future is extremely broad and deep.

Nor is the support for nuclear power restricted to governments. There is a handful of left-wing journalists, from George Monbiot to Leigh Phillips, who are emphatically pro-nuclear.[7] Not as an alternative to renewable energy, but as a supplement to it. As a supplier of baseload power, to back up intermittent energy sources like wind and solar. To quickly and expediently decarbonize the grid so that our new electric cars, buses and trains, and our new electric heating systems, don't accelerate the crisis. There being no zero-carbon options, nuclear is held to be almost as close as wind and water to that goal.

This doesn't mean they necessarily sign up to every public–private nuclear boondoggle. Monbiot was, for example, fiercely critical of the hugely costly Hinkley Point C project in a *Guardian* piece in September 2016.[8] But he defends the principle of a massive expansion of nuclear power, led by the public sector, as a safe and carbon-efficient part of the solution.

III.

The surprising thing about nuclear advocacy is that it comes at a moment of crisis and threatened decline for the industry. Look how fast we get to 'peak oil' and 'peak uranium'.

It took 240 million years of high-pressure underground incubation of dead florae to make the oil reserves. It took about a hundred years to exploit half of that. The uranium deposits were made by processes older than the planet itself. It took half a century to exploit half of that.

On average, according to the physicist Michael Dittmar, only 50–70 per cent of uranium resources can be mined, depending on the situation. It is estimated that all the available ore will be gone in at most seventy years.[9] In Europe, uranium mining largely ended in the 1990s. In China, where uranium supplies are still plentiful, and where the government is investing in nuclear power on that basis, peak supply is expected to be reached by 2042. In a 2011 gathering of the World Resources Forum at Davos, Dittmar estimated that 'plateau production' on the current basis would last about ten years, after which there would be a likely decline. But production has to increase to sustain nuclear energy's role in decarbonization. Even assuming a sustainable supply, or new discoveries countervailing against scarcity, the increase in demand is likely to send prices soaring.[10]

The search is on for a technological miracle to solve this problem. Currently, governments are investing in potential ways to extract uranium from seawater.[11] This is a fascinating line of research in itself, which can be traced back to UK military-scientific efforts in the Cold War. But it remains at an experimental stage. There is little chance of this becoming a global commercial prospect any time soon. At present, there are hard limits to this energy. And given the fundamental unreliability of the industry's estimates of 'reasonable assured resources', these limits may be pressing sooner than we think. Even with the resources increasingly exhausted, nuclear energy currently supplies no more than 3 per cent of global energy. Global nuclear energy use has been falling since 1996, according to the 2017 *World Nuclear Industry Status Report*. Even if a massive increase in nuclear energy usage was possible, it wouldn't put a dent in the problem.

IV.

Any energy source imposes three types of cost: carbon cost; financial cost; and opportunity cost.

Any energy-infrastructure investment has to be long-term. This is the opportunity cost. It has to lock you in for generations. Once you build an infrastructure you create dependencies which last. It excludes options, by definition. This is true of fossil extraction industries, wind power, hydro, solar – and it is especially true of nuclear energy.

This is not just because of the lifetime of nuclear power plants. Hinkley Point C is expected to have a lifetime of about sixty years, for example. But, rather, it has to do with the question of waste storage. The half-life of plutonium waste is 24,000 years, and it will be far longer before it is safe for the environment. Even though most existing reactors are based in rich OECD nations, where there exist infrastructures that adhere to minimal industry standards of 'best practice', the industry doesn't know how to store waste safely and securely for anything like that length of time. This would be true even if the industry wasn't, as Monbiot put it, a bunch of 'corner-cutting scumbags'.[12]

Take, for example, the San Onofre power plant in California. Southern California Edison, the firm operating it, built unsafe reactors and operated them outside of allowable limits for pressure and temperature. After a resulting radiation leak, the plant was shut down. The same people who designed the plant were given control of waste management. They were permitted by local authorities to use the site itself as the location for the burial of the waste.[13] They were given considerable latitude to bypass regulations requiring them to notify the public of ongoing issues. And the emergency preparation system was gutted, just for them.[14]

Legal action by residents forced them to move the waste away from a highly populated earthquake zone. But the container system they proposed remained the same: underground storage in dry casket containers made of steel and concrete. This is designed to last for about sixty years, but only guaranteed to last twenty-five years. And this is the solution used for most of the over 90,000 tonnes of nuclear waste generated by US power plants.[15]

In the UK, most waste has historically been stored at Sellafield. Because of the extremely dangerous (and desirable, for dangerous people) nature of the materials stored there, Sellafield acquired its own police force and fire brigade. But that didn't prevent a range of corner-cutting measures, such as habitual understaffing and storing nuclear waste in plastic bottles.[16]

The current ways of dealing with nuclear waste display extremely short-term thinking about a problem that demands 'deep time' thinking.[17] Managing nuclear waste will require future generations to engage in long-term investments and strategies to pay for the generation of energy today. That's another form of opportunity cost.

v.

This is, of course, linked to the issue of financial cost. As the *Economist* shows us, how you estimate the cost of any energy source depends on such factors as under what system and in whose interests the energy is produced.

Currently, the dominant strategy of the far-sighted elements in the fossil states is to try to engineer a situation in which renewables are commercially viable: for profit, in a capitalist market. For example, the UK government has a small pot of subsidy cash available for low-carbon energy sources, for which wind, solar, hydro and nuclear firms compete.

There are all sorts of perverse incentives for the UK government to favour nuclear in this competition, despite the fact that it has never been profitable without massive subsidy. The UK's civil nuclear industry has its origins in the military production of nuclear weapons.

This is why it is dedicated to reprocessing nuclear waste, to extract plutonium for new weapons production. It is also how it ended up alighting on the happy idea of using radioactive waste, depleted uranium, as a weapon.[18] So it is unsurprising that the UK government has chosen a white elephant in the form of Hinkley Point C as its energy flagship for the foreseeable future. The same perverse incentives apply in WMD-rich France, where the grid is overwhelmingly nuclear.

In principle, it should be possible for any government of the radical left to adopt a very different costing framework. For example, Labour's commitment to a publicly owned, decentralized energy system would change the calculus substantially. Profitability would not have to decide whether and how fast we move to renewables. However, such a government would have to give up the perverse incentives favouring nuclear energy. It's not clear that Labour would do that. John McDonnell currently claims to support the government's nuclear boondoggle.[19] Labour's 2017 manifesto committed it to 60 per cent renewables or zero-carbon resources by 2030, a form of wording that keeps the door open to nuclear energy. That's clearly in part a by-product of Labour's acceptance of the increasingly preposterous Trident system.

The problem is that this nexus of military, science and industry, and the growing role of military power in organizing climate mitigation – what Geoff Mann and Joel Wainwright call the 'climate Leviathan'[20] – makes it nearly impossible to base policy on a proper cost-benefit analysis. This is one reason why the CCC has been so pro-nuclear. It rests its case partly on findings that nuclear energy would be cheaper than renewables, a condition that, as the environmentalist Jonathon Porritt pointed out in a polemical exchange with Monbiot in the *Guardian*, applied only if the renewable infrastructure was not properly developed to make it cost-efficient.[21]

Indeed, as the 2009 *World Nuclear Industry Status Report* pointed out, the real costs of nuclear energy are often obscured for a number of reasons. Public contributions are often left out entirely from cost assessments, for example. But there are also, as suggested by the above, a range of valid uncertainties about the future availability of raw materials, as well as about the reliability of industry guarantees.

VI.

What, then, is the carbon cost of nuclear? And are we able to pay it? Carbon cost has not been priced by capitalist markets. Indeed, they don't have any idea how to price it. The whole point of such costs is that they are externalities.

More than that, the cost of carbon is not commensurable with market pricing. There is no way to quantify it in relation to the labour costs congealed in a particular good. It is a matter of the destiny of the whole human species. It is a political issue. This is why attempts to use market mechanisms to simulate a carbon price (say, carbon taxes or cap and trade) have been completely ineffectual.

The good news is that governments, with major exceptions, now accept the goal of zero net carbon emissions by 2050. That commitment is vastly greater than their actual preparations. Moreover, they seem to be hoping that fossil fuel giants will slowly melt away, or diversify into renewables, when they have massive investments in fossil energy that have to pay off for decades to be profitable. But it is worth having the target set down. The question is, can nuclear energy help get us there? Is it a useful bridge to zero net carbon emissions? What is the nuclear carbon footprint?

The IPCC is, again, strikingly optimistic about this.[22] It estimates how much carbon dioxide each energy source emits in an hour of use. Since carbon dioxide isn't the only warming gas, it also takes into account equivalents like methane. The measurement is called 'grams of carbon dioxide equivalent per kilowatt-hour' (gCO_2/kWh). And it suggests that nuclear energy is not just as good as renewables on this score but, on the balance of research, better than hydropower, better than solar, and about equal to wind power.

Of course, the IPCC does not do its own research. It cites the research being carried out by others. With regard to nuclear power, Annexe II of the Fifth

Assessment makes it clear that the figures are based on two meta-studies. These were conducted by Manfred Lenzen (2008) and Ethan Warner and Gavin Heath (2012).[23] This is a far smaller range of sources than those cited for any other energy resource, but it can be justified by the fact that they are meta-reviews, not single studies.

Nonetheless, just as their overall analysis omitted research critical of nuclear power, in their calculations the IPCC omitted Benjamin Sovacool's (2008) meta-review, which gave a much higher figure for nuclear carbon emissions.[24] The IPCC's median estimate of the amount of carbon dioxide emitted each hour by nuclear energy over its life cycle is $12 \text{ gCO}_2/\text{kWh}$. Sovacool's average estimate is $66 \text{ gCO}_2/\text{kWh}$. That would seem to place it well above geothermal, hydropower, wind and solar, except that Sovacool's average is a mean, whereas the IPCC are using a median. They are just claiming that half of the estimates of nuclear emissions their meta-reviews looked at were below the figure of $12 \text{ gCO}_2/\text{kWh}$. When the spread of assessments varies widely, a mean average is arguably more useful, depending on the reasons for the spread.

On that very point, as the aforementioned physicist Keith Barnham pointed out in the *Ecologist* in February 2015, the Warner and Heath survey poses various difficulties. Its meta-review takes into account ninety-nine estimates, which it treats as 'independent'. But these come from just twenty-seven papers, which means that studies giving a large number of different estimates for the same model are weighted disproportionately. As Barnham puts it, 'these are mainly analyses which report low carbon footprints'. Their methodology weighs down their median by inappropriately inflating the number of estimates that fall at the lower end. Beyond this, both the Sovacool and the Warner and Heath surveys include studies which overlook big parts of the nuclear life cycle.

VII.

Why is it so hard to figure out the carbon cost of nuclear energy? Along with the combination of perverse incentives, interests and ideology, one has to factor in difficulties that are intrinsic to the terrain.

A great deal depends, for example, on such factors as the quality of the ore being extracted. The lower the quality, the more carbon is emitted during its production. At the lowest quality, 0.005 per cent concentration, the carbon emitted is higher than for natural gas, according to a 2011 life-cycle analysis of nuclear power by the Austrian Institute of Ecology and the Austrian Energy Agency.[25] There are variations in the location and method of extraction (open-pit, underground, *in situ* leaching) that can make a difference.

Much besides depends on what assumptions you make about the inputs associated with the life cycle of the energy supply. Different surveys of nuclear energy emissions make wildly different assumptions about reactor construction, operation, fuel preparation, decommissioning and waste disposal. Those, naturally, will be inflected by unconscious ideological assumptions. The assumed energy context also matters. At the moment, for example, the energy required to produce even renewable supplies would require significant carbon emissions. That's because most of our energy still comes from fossil fuels. If, as in Norway, 99 per cent of our electricity production came from renewables, then the estimate would be far lower.

It is also not simple to compare like with like. Nuclear energy has been the object of government strategies, funding and infrastructure building for decades. Any uncertainties that remain are there to stay, and indeed may get worse as stocks are exhausted. After all, the carbon cost of extracting and processing fuel is one of the major factors in the overall emissions of nuclear energy. The renewable energy sector doesn't have any fuel factor to deal with. It is, by definition, renewable. Its sources are supplied, abundantly, by the natural world if they can be harnessed. But it is also underdeveloped, a source of energy for which infrastructures have not been built. There naturally remain uncertainties, depending on where the wind farms or dams or solar panels are situated, and the intermittencies of source to which they are subjected.

Nonetheless, what is clear is that the atomic genie is not the clean, efficient, unproblematic source that its apologists pretend it is. The reasons given for claiming that nuclear can do the job of renewables, only more stably, are not the good reasons they imagine. The solution appeals to those on the 'I Fucking Love Science' left, because it seems to avoid tragedy. A massive statist expansion of nuclear energy will solve our problem almost overnight, so we really don't face too many hard choices about consumption and fundamental economic restructuring: it's just a matter of political will.

This evasion of tragedy is at its worst in Monbiot's claim that one doesn't have to choose between nuclear and renewables.[26] You unfortunately do. The two are very different kinds of infrastructure, linked to very different types of statist logic, and a very different calculus of human survival (or, where nuclear weapons are involved, extinction). And every pound you spend on nuclear energy is a pound not spent on developing the essential renewables infrastructure. And that is a choice that might just end up costing the planet.

POSTSCRIPT

February 2022

This essay is about nuclear fission and the military-industrial complex promoting it. Thus far, our only access to energy produced by nuclear fusion has been from the sun. But what if a small star, a miniature version of the sun, could be safely produced on earth? What if atoms could be fused in a star machine, and the energy from it harnessed? In principle, it would provide abundant, low-carbon, low-radiation energy. In February 2022, there was much excitement about a breakthrough in this field achieved by scientists working at the JET laboratory in Abingdon, Oxfordshire. In their experiment, fusing two forms of hydrogen, they produced eleven MW of power over five seconds. That's roughly 11,000 kWh, or enough to boil about sixty kettles of water. However, the fusion reaction was initiated by two 500 MW flywheels, meaning that the result was a net deficit. This might be overcome when energy is produced at sufficient scale, but that is decades away. And even then, its safety and low-carbon emissions would need to be demonstrated in practice. The problem is that the energy transition has to begin now, and it has to create infrastructures that endure. Nuclear fusion probably isn't going to be much help with the transition.

WHAT, OR WHOM, WILL WE EAT?

26 JULY 2019

Every now and again, a headline just begs to be interrogated. 'Cheap food', it is reported, is destroying both human and ecological health. It's not untrue. In fact, it's an urgent and under-studied fact.[1]

But, you might think, what's the alternative? Expensive food? Smaller diets? There seems no way to impose such strictures only on the rich, without actually hurting the poor. Moreover, if we insist that we can't feed people cheaply without causing ecological collapse, do we have to become Malthusians? We can't evade the fact of natural limits to the planet's systems. But how close we are to those limits, and the effects they have, is obviously contingent on our social system. And, of course, the way it uses technologies.

There is, for example, no necessary reason why two billion should go hungry or malnourished when we're producing more food than ever in human history.[2] Nor is it inevitable that a planet of 11 billion people should entail massive deprivation, even though many catastrophic ecological thresholds will have been crossed by then. In principle, there must be some number of humans that would simply be untenable for the planet, no matter what. But at least as regards food consumption, we're not there.[3] So if cheap food is producing an ecological crisis, it is a matter of energy conservation on a planetary scale.

Food is our indispensable energy source. As a capitalist product, it is at the intersection of several energetic systems. The 'web of life' is one such, circulating enormous quantities of chemical and caloric energy, mostly drawn from the sun and inorganic chemical compounds. The first stage of production leading to the food on our plates is not the planting of seeds or rearing of cattle, but the autotrophic production of chemical energy known as photosynthesis and chemosynthesis.

Human labour is another energetic system, emerging from the latter, distinguished by the fact that it is capable of being organized by collective symbolic intelligence. Capitalist industry, with its technoscientific organization of life processes, is another. Fossil capital, the infrastructures and energy flows specifically organized around the exhumation of ancient remains of life, to enable specifically capitalist production, is another. Every bite we eat is a congealed quantity of these several forms of energy.

The energy inefficiencies of this food system are, by now, well understood. The food we cultivate consumes many times more calories of energy than we actually eat. Beef, for example. A mature steer consumes thousands of pounds of corn and soy feed in its lifetime, and its slaughter produces about 500 pounds of beef. We're making enormous withdrawals from the energy system with every burger. It goes without saying that long supply chains, in addition to contributing to the waste of about a third of all food, also add to the energy cost in the form of burnt fossil fuels. But even if we're vegetarians who eat strictly local produce, Raj Patel and Jason W. Moore point out in *A History of the World in Seven Cheap Things*, the massive use of fertilizer to increase food production means that it takes ten calories of oil to produce a single calorie of food. And that food is of diminishing nutritional value, beyond its crude bulk of calories.[4]

The irony of the system of 'cheap food' is that it is hugely expensive. It doesn't register as such within the framework of capitalism, as Moore points out, only because the amount of wage labour necessary to produce the food is relatively low. And the wage bill is further suppressed by forms of forced labour, and dispossession (which tends to make more people available for low-cost labour), just as it was historically suppressed by slavery and New World genocide.[5] Lowering the cost of food in turn lowers the composition of costs that make up the average price of labour power. It makes work cheaper for capital to buy. (It is not, in this sense, Marxists who have a 'labour theory of value', but capitalists: Marxists simply raise this precept from unconscious to conscious awareness.)

But 'cheap food' is not just expensive in the sense of being wasteful, taxing, an unsustainable burden on the earth's energetic systems. It is also quite visibly destructive of the energy systems on which it relies.[6] By decomposing and degrading the web of life, killing off species, it puts out of circulation enormous quantities of energy. By carbonizing the air and water, and acidifying the oceans, it kills off the phytoplankton responsible for most primary production and also for about half of the oxygen – which, you may know, cellular life dependent on aerobic metabolism needs, as part of a chemical process upstream of the Krebs cycle, to produce energy. This destructive force Moore calls 'negative value'.[7] This is the most perverse system of geoengineering in human history.

This system of 'cheap food' is reaching hard limits. It is possible that capitalist technoscience will come to its rescue. The would-be pioneers of the next 'green revolution' are coming from outside conventional agriculture, in the form of tech giants like Microsoft and Google. Digitized food production is offered as a way to reduce factory waste. High-yield farming promises to spare natural habitats.[8] Yet the rescue would only be momentary. The paradox of energy efficiency, the 'Jevons paradox', is well known. Such techno-fixes, leaving capitalist relations of production in place, will simply enable the more intensive exploitation and destruction of the earth's systems. They will plunder that much more energy.

The problem isn't the technologies per se, but the human purposes they automate. There is no necessary reason why high-yield farming should result in more land use, more energy use, more species destruction. That's just the way it works under a system of anarchic competition, where the only imperative is to accumulate profit. One could imagine, with a different set of imperatives, a degree of global cooperation and planning, and a renewable energy infrastructure, that such techniques as high-yield farming, aeroponic warehouses and so on would be of use in a half-earth strategy.[9]

The problem with the 'cheap food' system is that it is only 'cheap' for capital: it really isn't remotely cheap for most of the world's populations of people, animals and plants. It is in fact enormously expensive, and we are beginning to pick up the tab.

A NOTE ON
CLIMATE SADISM

22 AUGUST 2019

Think about middle-aged journalists and right-wing politicos tweeting at Greta Thunberg to inform her that they've just booked a long-haul flight with zero guilt, or would be quite pleased to see her have a fatal accident.

Or think about right-wing men buying SUVs to 'own the libs'. Or soul-dead contrarians relishing the latest heatwave as a great chance for a lovely day at the beach. Or Trump responding to the perturbation of the polar vortex, induced by global warming, by wishing for a bit of 'good old-fashioned' global warming.

Now think, more gravely, about Bolsonaro calling indigenous people 'animals', threatening to crush them in the interests of capitalist development, escalating deforestation.[1] Cheerfully, ebulliently trashing one of the earth's last survival systems while unleashing racist violence not seen since the days of Brazil's military dictatorship.[2] And then, when ranchers take advantage of the political mood music to set fire to the reservations[3] and precipitate the worst rainforest fires in history, blaming the NGOs for it.[4] Far-right trolling is always very meta.

There's something profoundly strange about climate sadism. It usually starts with an assertion of ruthless, competitive, indifferent self-interest. But it is far from really being indifferent. It is, on the contrary, both paranoid (climate change is a scam, the Chinese are mugging us off) and deeply invested in the reaction that it provokes (ha ha, carelord tears are delicious). This is a good reason to pay close attention whenever anyone talks about 'self-interest'. As Freud suggested, there is no self without others, no self-interest that is not an interest in others, no individual psychology which is not also a mass psychology, and no 'I' which is not peopled by everyone else in one's life. Everyone contains masses, love objects and hate objects, and all 'self-interest' is wrapped up in passionate identifications, with all the ambivalence and aggressivity that comes with that terrain. And when it comes to sadism, as Lacan argued, it is always enacted for the other's gratification. Trolls don't have to enjoy what they do, because they are instruments of the greater pleasure of their collective.

With climate sadism, contrary to what we usually hear, assertions of brutal self-interest are a decoy. What is at stake is passionate libidinal attachment to the group, and its collective loves and hates. And the group cathexis is organized by an ideal: let's say, the ideal of accumulation without limits, petromodernity for eternity. And since denial never destroys the truth, but merely represses it, this is often subtended by a dark, supremacist fantasy: even as everything goes to hell, racial and national dominance will secure the future for those quick and ruthless enough to make hay while the sun burns ever brighter.

And yet. There is another indestructible truth of the situation. No matter how brutally the resources of the planet are marshalled to protect racial and national boundaries, the collapse of the earth's systems would not respect those boundaries. We face the prospect of a collapse of the food chain and the exhaustion of fertile land, the depletion of oxygen as marine life is killed off, the flooding of major cities as sea levels rise, unliveable temperatures and extreme weather making large parts of the world uninhabitable. A few might be spared severe hardship and danger, but this will not be a pleasant world. The troll's logic is: 'It's funny that you people care whether we live or die, so I'll enjoy the look on your face as I destroy the future of all, including anyone I have ever loved, and even my big stupid self, and laugh about it.' Climate sadists are masochists in denial.

THE SUN NOW
EMBRACES NATURE

9 JANUARY 2020

We inhabit a planet that, for better or worse, is regularly on fire. The oxygen-rich atmosphere. Soils packed with organic fuels. Air crackling with lightning. Flammable carbon-based organisms. All conspire to create conflagrations.

Wildfires erupt in the forests of Portugal, in the taiga, in the Australian bush, in the woodlands of California and in the Amazon. Some of them have burned since the Ice Age. They would do so even without human intervention. From south to north, west to east, earth scientist Stephen Pyne tells us, every ecosystem has its own fire regime.[1] They have evolved to burn. Whole living systems depend on fire.

We, too, have evolved to burn. Human beings became the apex predator on the planet in part through our control of life-giving fire. There is a way of talking about this which dissolves the history of fossil capital into a general human propensity to burn. It becomes, at worst, a story of the hateful folly or rapacity of human beings. Yet the energetic basis of human civilization does matter, and fire has historically been at the centre of it. Pastoral fire is good husbandry, the fire stick a farming implement. Controlling fire gave early hominids access to quantities of energy inaccessible to other species. It gave them access to everything that could be cultivated and eaten on land. It made edible that which would otherwise kill us.

We can do without burning the fossil remains of ancient trees and mosses, but we couldn't be without fire. Too little is as bad as too much. Without fire, especially in lands poor in nutrients, where the burning off of old growth resets the biological clock, some biotas would die off. When fire is suppressed in some parts of the world, the resulting damage to the local ecosystem makes uncontrollable wildfire a more likely occurrence.

This, famously, is geographer Mike Davis's 'case for letting Malibu burn'.[2] The grasslands of California burn, on average, every couple of years, with no great loss as far as the plant life is concerned. The shrubland burns once every five years. The chaparral and woodland communities fringing Malibu currently burn once a decade. This is probably a more frequent rate of conflagration than it would be were it not for climate change, and other manifestations of human intervention. Nonetheless, it is futile, wasteful, reckless, to build rich communities in these flammable coastal areas when they are destined to go up in smoke. Fire suppression efforts change the biochemical composition and moisture resistance of the soil and make the fires more extreme.

Australia, likewise, is meant to burn. Currently, around 5 per cent of the Australian land surface is burned by wildfires each year. This destroys 10 per cent of the continent's 'net primary productivity' – that is, the ability of its tropical forests, woodlands and savannahs to photosynthesize solar energy. Again, these fires are more severe than they would be without climate change. But if they didn't happen at all, parts of the continent would die. Since Australia broke off from ancient Gondwana, it has evolved ecosystems packed with pyrophytic and pyrophilic trees and plants, across large, low-nutrient, parched landscapes.

For millennia, aboriginal farmers have used fire to cultivate a vast expanse of land. Before colonization, this was the means by which lush, overgrown forests were 'tidied up', and wildfires contained. Similar forms of pastoral burning were used by Native Americans, creating a mosaic, constantly transformed landscape which colonists mistakenly assumed was pristine and undisturbed by human activity. John Locke's justifications for conquest, enslavement and elimination were, of course, partly predicated on the claim that Native Americans hadn't mixed their labour with the soil. The same practices were found by French colonial travellers in Nigeria, Benin and Senegal, who originally deployed the term 'fire regime' to describe such indigenous fire management as a regressive, unnatural disaster being visited on nature by idiotic natives.

Colonists weren't invariably so obtuse. Coming from European societies, they were conditioned to think that the fire ecologies obtaining in the temperate core of Europe – not the boreal forests, or the woodlands of southern Europe – were normal and correct. Yet botanists, foresters and even some settler colonies understood indigenous fire management. They used the term 'fire regime' to refer to natural cycles which prudent human intervention could trim and tailor. But while that understanding was eventually generalized, it didn't prevent the extensive extirpation of these fire management techniques in favour of the outright suppression of wildfire. The ironic effect of colonial fire suppression was to generate an increase in severe wildfires. This, as much as anything else, qualifies as what Alfred Crosby called 'ecological imperialism'.[3] The biological expansion of Europe entailed exporting its fire regime.

This domestication of fire was contiguous with, and logically continuous with, the rise of fossil capital and slash-and-burn deforestation. An expanding global capitalism sought to snuff out naturally occurring fires, while setting fire to fossil carbon and burning down forests, as part of the same drive to master the life process and subordinate it to the production of surplus value. As a result of the combined effects of all three, the ecology of fire is, everywhere, turning lethal. Fire regimes, already exacerbated by human activities, have been radicalized by global heating. The correlation between rising average global temperatures and the increasing severity of wildfires is obvious, as is the explanation.[4] Dozens of studies have predicted that higher temperatures would cause fire seasons to start earlier, finish later, become more severe and kill off important ecosystems.[5] This is what is happening. And it engenders a vicious feedback loop. The increasing incidence and severity of fires accelerates global heating, both by pumping out more black carbon into the atmosphere, and by speeding up Arctic melt and reducing the ice's albedo effect. A warmer climate means yet more deadly wildfires.[6]

In 2018 California saw its deadliest and most severe wildfires, which killed over a hundred people and consumed 766,000 hectares. In Australia, 6.3 million hectares of land have burned since September, in the worst wildfires the country has experienced. The boreal forests of northern Russia are burning more often, and for longer. Last year's Siberian wildfires were horrifying, destroying three million hectares – 'unprecedented', according to Mark Parrington of the Copernicus Atmosphere Monitoring Service.[7] In fact, earlier fire seasons triggered by climate anomalies in Russia were even worse: in 1998, 2002 and 2003, respectively, 6.9, 7.5 and 14.5 million hectares of forest and surrounding areas were burned. In the 1998 fires alone, 516 million metric tonnes of carbon dioxide and 500 million metric tonnes of carbon monoxide were emitted.[8] In Amazonia, it is impossible to isolate the effects of climate change given the extent of deliberate burning for the purposes of deforestation. How much of last year's surge in August fires was triggered by global heating, and how much by the election of Bolsonaro, and his promise to liberate capital and wage war on indigenous communities?

Many of us could, cynically, turn off the news and forget about this, if only the fire would stay where it was. But fire never does. The fact that we have an atmosphere means that every fire is a global event. Wind and vapour transport gases, aerosols and particles of black carbon, concentrate and distribute them. The wildfires of California and Australia, and the slash-and-burn deforestation of Amazonia and Indonesia, are not far-off happenings. They are happening, differently, and over different timescales, to everyone. The fires in New South Wales and the floods in South Yorkshire, for instance, are part of the same story. This isn't a story about 'them', I mean to say, but about 'us'. It will inevitably come for us.

The current fire regime is almost designed to spiral furiously. Pyne argues that because of it, we're entering a 'pyrocene': a 'planetary fire age', 'like an ice age for fire'.[9] And while fire is inevitable, human beings are not. The planetary conditions which enabled us hominids to emerge and evolve to our current status are contingent, a deep-historical fluke. We have no inherent right to survive, let alone dominate. And no world-historical spirit, or deity, is going to step in and redeem the species if we insist on burning ourselves out of our own planetary home.

THE DARK SIDE OF
CARBON DEMOCRACY

21 JANUARY 2020

Greta Thunberg addressed Davos earlier today. To explain, once more, that 'our house is on fire'. To complain, once more, that every time this is explained to our leaders, they do nothing. To demand, once more, that they act to stop the crisis.[1]

Why does nothing happen? What is the obstacle? It isn't a lack of knowledge. It isn't a lack of consensus. It isn't a lack of public support for the principle of fighting climate change. Most of the carbon emissions that have taken place in the history of humanity, according to Wallace-Wells, have occurred when the climate consensus was robust, acknowledged by most governments, and the subject of formal international agreements.

In recent years, there's been a trend to blame the stalemate on democracy. This has always been a significant thread in ecological thought, from Garrett Hardin's 'The Tragedy of the Commons' to William Ophuls's 'Leviathan or Oblivion?' More recently, ecologists David Shearman and Joseph Wayne Smith argue forthrightly in *The Climate Change Challenge and the Failure of Democracy* that ecological threats could be traced directly to 'common threads in the functioning of democracy'. 'Human nature' is poorly designed to face up to imminent threats, dismissing them as readily as the human mind dismisses death. 'Authoritarianism,' they argue, 'is the natural state of humanity.' The best outcome would be a 'Platonic form of authoritarianism based on the rule of scientists'.[2]

However, well beyond the ranks of ecological authoritarians, now extending to liberals, there is a growing chorus of concern that democracy is getting in the way of fixing the problem. Nathaniel Rich's famous *New York Times* piece 'Losing Earth',[3] for example, draws on the doleful species-indictment of the eco-authoritarians in concluding that 'we', 'us', the demos, are to blame. He is, however, far too complacent, far too solemnly unruffled and un-panicked, to be legitimately honoured with Sarah Jaffe's label, 'gloomdude' (don't knock it 'til you've tried it). There is a more procedural critique of democracy, advanced by David Runciman in *Foreign Policy* in July 2019, and Edward Luce in the *Financial Times* in January 2020. To wit, they argue that politicians need to take measures that will take generations to bear fruit, and even then go unrecognized. Electoral cycles prohibit this kind of far-sightedness, and party polarization undermines the necessary consensus-building. But whereas Runciman actually suggests that this requires a more radical, participatory democracy, Luce can only revert to the classically liberal strategy of appealing to economic self-interest: look how expensive the wildfires are.

Far more bleakly, Simon Kuper of the *Financial Times* aims in an October 2019 essay to prove that there really is no conveniently democratic solution to climate change.[4] He argues that green growth, even of the Green New Deal variety, is predicated on an unavailing 'win-win' logic. In the real world, growth means emissions. The amount of carbon required to produce $1 of GDP would need to fall ten times faster than it is, if we are to feed a growing population and avoid disaster. The only solution is a long economic depression. And since no electorate will vote to decimate its own lifestyle, 'we'll never find out' if democracy could survive a post-carbon world.

The worst thing to do, faced with such alarm, is to dismiss it outright. If there is a glut of doom-mongers, it is because the market in doom is booming. There are excellent reasons for scepticism, of course. The idea of a benign climate dictatorship is a fantasy that will never get off the ground, there being no social base for it. Authoritarian regimes tend to have little interest in saving the planet, and the degree of social change entailed by it requires the sort of public buy-in that only democracy can assure.

The analyses blaming climate stalemate on democracy aren't without evidence, of course. Look at Trump in the US, Scott Morrison in Australia, Bolsonaro in Brazil. Look at the gilets jaunes, and the Ecuador protests against fuel subsidy cuts. But Trump lost the popular vote in 2016, the 'climate' vote was split in Australia, and in Brazil the entire political establishment had just collapsed before Bolsonaro was elected. Even when their rivals were in office, moreover, they delivered very little on the climate crisis. Protests like the gilets jaunes will tend to occur when there's an obvious social injustice, detached from any serious effort to decarbonize. It isn't as though the Ecuadorian president Lenín Moreno was breaking with extractivism. It also doesn't quite seem the right time to claim that millions of people would never vote to reduce their own living standards. Ethno-nationalists seem to have little trouble in persuading millions to give up the eternal mantra of 'jobs'.

Further, as Alyssa Battistoni and Jedediah Britton-Purdy point out in their article 'After Carbon Democracy', none of these authors has anything to say about capitalism. This is a huge omission, as if one was to try and explain ethnic cleansing without mentioning the nation state. How can one seriously prescribe economic depression as the answer, blame the public for opposing that, and say nothing about whether any capitalist class in the world would tolerate deliberate anti-growth policies? Blaming democracy and 'human nature', they say, can merely rationalize a refusal to face the problem and act.[5]

Still. And yet. There is a 'there' there. Democracy is not 'to blame'. And it is difficult to see how any viable, lasting solution can be found without some form of participatory democracy. However, this does not mean that democracy as we have known it is not implicated. Nor does it make democracy the eternal ally against the apocalypse. Nor does it help us answer the question of what kind of democracy could survive and allow us to survive.

Modern capitalist democracy, Timothy Mitchell tells us in *Carbon Democracy*, rests on fossil fuels.[6] Democracy as we know it emerged thanks to powerful working-class movements, built around coal extraction. The shift to oil modulated the administration of democracy. Eroding working-class power, it left in place a form of democratic politics in which consent was secured by the promise of plenty: 'a limitless horizon of growth'. The organized working class has now been decimated for decades. Limitless growth is over. The era in which capitalism could coexist with mass democracy may be ending, and there is no evidence that sufficient democratic forces exist to challenge capitalism. We have no right to assume, given the crisis of democracy, that it will persist without being revolutionized.

The crisis of democracy is obviously not unrelated to the rise of disaster nationalism, which is strongly tied to denihilist politics. As Michael Mann argues in *The Dark Side of Democracy*, democracy is susceptible to brutal torsions wherein the demos is conflated with the ethnos.[7] Particularly in crisis conditions, core features of democracy lend themselves to ethno-nationalist fury, in which class-like antagonism is transposed into a zero-sum ethnic struggle. The modern curse of ethnic cleansing and genocide is a curse of unstable democratic regimes. Now consider the scarcities, famines, disasters that the climate crisis will give rise to, and the effects this will have on our already shaky democratic systems.

We would like to think that the visible reality of an onrushing threat would destroy denial. At this point in history, we know better. If dealing with the climate crisis appears to threaten the way of life of millions, then it is just as likely to inflame denialism and its authoritarian recourses. As outlined in the October 2019 issue of *Salvage*, denihilism is not the only option for reaction in these circumstances. From green nationalism to eco-fascism, there are ways in which both democracy and ecology can be turned against themselves, involuted, and allied to the species' war on itself.

What is at stake, for all of us, is not just the question of whether we get to vote for representatives. It is not even whether we can convoke a 'thicker' and more participatory democracy in time to address the crisis. It is the question of civilization itself. This is the political and cultural problem, the threat of collapse, that in another era Keynes was trying to address through economics. This was the question that Rosa Luxemburg's so-called 'Junius Pamphlet' addressed during the thundering massacre of the First World War: socialism or barbarism?[8] What would a viable, post-carbon civilization look like? What kind of democracy would be equal to the depth of the challenge we now face, and how do we get it? We shouldn't act as though we know the answers to these questions.

WHAT IS AN IDEOLOGY
WITHOUT A SPACE?

9 JULY 2020

How long can fascism remain in denial? There have been a number of stories lately about rising eco-fascism. Indeed, the far right's adhesion to the climate-denial industry was by no means inevitable, and it won't necessarily continue that way.[1]

For the last few decades, the line has been: it's not real, it's a globalist conspiracy, they want to crush national sovereignty and give our wealth away to Third World moochers. Even if it is real, bring it on. More sunshine, warmer weather, what's not to like? And, sotto voce, if it kills off the weak, so much the better. Yet this is just not tenable.

There are growing signs of a far-right effort to articulate some version of environmentalism. The struggle for them has not been the absence of an ecofascist tradition, to which I'll return in a moment. Rather, since the far right thrives on resentment and collective hate, it needs to find the right friend/enemy distinction. That is beginning to happen. Look at German ecofascists saying: 'Let's chase the globalists off our acres.'[2] Look at the Fox News host Tucker Carlson saying: 'Isn't crowding your country the fastest way to despoil it, to pollute it?'[3]

Look at the altright.com website portraying nature as savaged by Jewish 'Unnatur': 'a modernist, capitalist, classically Liberal, materialist Jewish conception

[...] wiping out nature all over the Earth for the sake of higher profit margins'.[4] Look at French neo-fascist Marine Le Pen claiming that 'nomadic' immigrants 'do not care about the environment' because they 'have no homeland'.[5]

According to this fantasy, nature is white property. It is beauty, open space, the white wilderness of settler-colonial imagination, or the völkisch forests of European Romantic nationalism. It is the *Heimat*, the homeland. And it is threatened, in this scenario, by a coalition between feckless, rootless moochers and a dark conspiracy of global, rootless string-pullers.

The fact that fascism is most strongly associated with denihilism today, rather than environmentalism, is a result of certain contingencies. After the massacres in Christchurch and El Paso, however, the history of völkisch nationalism, counter-Enlightenment thought, misanthropic racism, fascism and settler colonialism in environmentalism has become more widely known.[6] And, for British readers, this is not just a story about elsewhere: say, about Ernst Haeckel, Savitri Devi, Alain de Benoist, Renaud Camus, Garrett Hardin, Hervé Juvin, Björn Höcke and Dave Foreman. It includes Jorian Jenks, fascist, Oswald Mosley ally, pioneer of environmentalism and the organic movement, and co-founder of Lady Balfour's Soil Association. It is a tradition that the British far-right, from John Bean to John Tyndall, has frequently sought to cultivate.[7] And some version of this well-developed ecofascist structure of feeling arguably inflects the green nationalism of English writers like Paul Kingsnorth, whose disdain for 'globalism', the 'global distancing of humanity from the rest of nature' and search for 'ecological Englishness' is congruent with his unease about immigration. 'English people had become an ethnic minority' in cities like London, Kingsnorth complains, leaving 'growing numbers' of English people 'beginning to feel unmoored and unspoken-for'.[8] It may also have something to do with some of the worrying 'blind spots' about racism that have appeared in the UK environmentalist movement lately.[9]

That this sensibility is not dominant today derives, as far as I can tell, from two sources. First, since the 1970s the ecological emphasis has been forcibly shifted by emerging science from land and nature conservation to global heating. Conservation is generally about protecting the local. In one form of ecofascist idiom, what one conserves is the habitat of the ethnos. The land, from that point of view, is where nature and culture meet in a form of life that must be protected from 'invader' species and people. Global heating, and associated problems like ocean acidification, species extinction and now pandemics, can only be apprehended in terms of a planetary consciousness. Not only that, but they can only be solved by cross-border cooperation. An upcoming book from Andreas Malm and the Zetkin Collective[10] will show that the far right is particularly revolted by the idea of an energy system that isn't tied to land and territory. Wind! Sunshine! How can you claim national property in those goods? Second, the left won control of the environmental movement in a series of decisive struggles that took place in the 1980s and 1990s. In doing so, it marginalized the nationalists, anti-immigrants and Malthusians. Earth First! founder Dave Foreman was among the eco-Malthusians. In the 1980s, he had argued that immigration to the US should cease, that welfare should be denied anyone with more than two children, and that the state should limit women to one child. As Ethiopia was struck with famine, he argued that aid would be misguided and that 'the best thing would be to just let nature seek its own balance, to let the people just starve there'. He eventually left Earth First! in the early 1990s, when most of its members rejected his views: it had succumbed to 'pressure and infiltration from the class-struggle/social-justice left', abandoning 'biocentrism in favour of humanism'.[11] It became too concerned with environmental racism, the poisoning of workers and the pollution of indigenous communities.

Fascists could, were it viable, remain comfortably denihilist. They have no problem denying science and portraying the consensus as a monstrous conspiracy. However, that could only work for as long as the effects of climate change are remote, abstract or in the future. If climate change was something that was coming in a few decades, or only affected those designated as infrahumanity, fascists could continue to cheerfully deny. The more sensuously concrete and 'local' these effects, the more they rip unpredictably through rich societies, the more Australia burns, the UK floods and the US is blighted by storms, wildfires and ice, the less plausible this denial becomes. The conversation is going to move on, past denial and covert affirmation. And the age of pandemics will add urgency to renewed ecological thinking.

A struggle over the environmental future, over how we mitigate and adapt, who pays the price for inevitable reductions in our global food intake, how energy is stored and shared, what counts as wealth and who gets it, is unavoidable. It's difficult to see a non-violent, demo-cratic resolution of that struggle. It would be complacent not to expect the far right to stake a bigger claim in that struggle than it already has.

ULTIMA THULE: AN OBITUARY

17 DECEMBER 2020

I am wakened each dawn
Increasingly to fear
Sail-stiffening air,
The birdless sea.

PHILIP LARKIN,
The North Ship, '65° N'

So awfully still, with the silence that shall
one day reign when the earth again becomes
desolate and empty.

FRIDTJOF NANSEN, *Farthest North*

She came to the limits of the world, to the
deep-flowing Oceanus [...] shrouded in mist
and cloud.

HOMER, *The Odyssey*

In February 2015, I found myself, quite by chance, in the Arctic. Polar perturbation had brought the Arctic to me, in Toronto. It had frozen the Midwest and the East Coast of the US. Lake Michigan and the coast off Maine smoked like the black seas of the north. The air was mercifully calm, but the temperature was well below minus twenty. It cut through layers of clothing, skin, flesh and bone, right to the marrow. It seemed to freeze everything in my body, as it froze the flow of the ice sheets: brain, blood and biomechanics.

Hominids arrived in the Arctic during the late Upper Palaeolithic. The first hunter-gatherers arrived in the north of Europe and Russia as part of the out-migration from Africa as early (or late) as 40,000 years ago during one of the interglacial periods. They crossed the Arctic Circle around 30,000 years ago, and would have been repeatedly driven back by the advance of ice. But throughout the Bronze and Iron Ages, populations grew in Arctic Scandinavia and Russia.

The first recorded encounter with the Arctic was much later, in the fourth century BC. But the experience of those early explorers must have been something like that of the geographer Pytheas, when he travelled far north, to a strange place where the sun faintly shone at midnight. 'Beyond Thule,' he wrote, naming an island believed to be the northernmost point in the world, 'we meet with the sluggish and congealed sea.' In the Arctic, evoking the ice packs and dense sea smoke, 'neither sea nor air, but a mixture like sea-lung [...] binds everything together.'[1] Scorned by his peers, who preferred to think of the Arctic as a mythic place, the land of the dead, the abysmal chasm, the chaos at the origin of the world, here was the first heart-twinge of the polar sublime.

For as long as there have been human beings, the Arctic has been frozen. In planetary terms, this is scarcely a moment. As late as 75 million years ago, during the Cretaceous period, the lands of the far north were populated by tall redwoods, ferns and swamp cypresses, not to mention gliding lemurs and dinosaurs, while plesiosaurs and sharks swam the warm waters. Fifty million years ago, in the early Paleogene, when the circumpolar ocean was largely landlocked for the first time, primitive conifer forests would have been thronging with the dinosaurs' immediate descendants, birds.

On the cusp of the Quaternary period, about 2.7 million years ago, planetary temperatures began a sharp descent. Thus the primitive accumulation of Arctic ice. The drop in average temperature alone would not have been enough to cake the North Pole with its perennial ice cap. Rather, the icing was caused by a widening differential between summer and winter temperatures in the north. In the summer, higher temperatures meant more water was evaporated. In the winter, the evaporated water froze and fell as snow, while the colder temperatures froze the sea. Instead of gliding lemurs and birds, the far north was populated by woolly rhinoceroses and mastodons. The glaciations of the Quaternary period, caused by the slight eccentricity of the earth's orbit around the sun, added layers.

Crystal upon crystal, a dendritic network of networks, frazil, rapidly forming a greasy rime in which bobbed lumpy shuga, congealing into slushy pancakes, compressing, lithifying, metamorphosing, stressing, warping, squeezing out saline and air, forming bands of lucent blue ice and frosted bubbles. The density of it: 917 kilograms per square metre. The strange presence-absence of it: there was no consistency in colour or texture, the same icescape could be dull grey and foggy, or brilliant with ocean blue, jade green and simmering solar orange. It could be gothic and spectral, like Gustave Doré's illustrations to Samuel Taylor Coleridge's 'Rime of the Ancient Mariner', or spiritual and refulgent, like Lawren Harris's mystical northern uplands. It could be dirty and slightly phosphorous, as in Edward Wilson's Antarctica, as noir as Emil Schulthess's Antarctica, as atmospheric as Eric Ravilious's Arctic, as pristine as Eliot Porter's Antarctica, or as folk-horror as Andy Goldsworthy's Arctic. The light-bending treachery of it: the light waves were not so much absorbed as twisted round ice particles and refracted to produce optical inversions, phantom shapes, multiple suns (parhelia), illusory cliffs, towers and metropolises of ice (Fata Morgana). The haunting auditory hallucinations: explorers swore blind that they heard distinct human screams coming from the still ice. Dr Elisha Kent Kane, who travelled on the American First Grinnell Expedition in 1850, called it a 'necromantic juggle'.[2] The cover was never complete. Arctic winds or tides formed pitch-black smoking leads and polynyas – openings in the ice, along which Arctic whales, narwhals and walruses navigated. And the snow silently settling on the ice formed the firn from which new records, new deposits of geohistory would be comprised. 'A stratigraphy of snow', as Stephen Pyne puts it.[3]

This, the Arctic status quo for the entirety of human existence, was scarcely a promising condition for flourishing life. The solar energy feeding primary ocean production, and allowing plant growth, was wan and weak. The ice cap prevented the oceanic upwelling that would draw phosphates, nitrates and silicates from the benthic depths to the sunlit pelagic surface, to feed krill. On the snow-capped land, the soil was acidic and there weren't enough worms, fungi and beetles to decompose dead plant matter and enrich the earth with nutrients. The permafrost covering most of the Arctic land mass permeated to depths of up to five thousand feet in Siberia, and left much of the summer landscape boggy and plant-less. The modern Arctic ecosystem, which began to emerge about 12,000 years ago, at the end of the Pleistocene, nonetheless accommodated evolutionary wonders even within its extremely curtailed food chain, spanning from diatoms to bears.

In the spring and summer months, when the ice melted and calved, the Greenland ice cap birthing a thousand bergs – 'some as tall as cathedrals', as Darwin puts it in *The Voyage of the Beagle* – and the energy from the sun warmed the Arctic ocean, primary productivity blossomed. The underside of the pelagic ice floes, and the salty brine channels that had formed during freezing, hosted a complex ecosystem of single-cell animals, some with glass shells, which thrived on the solar rays and nourished copepods and amphipods, beginning the food chain that ended with polar bears. The perennial ice cover also preserved a constant climate in the dense, saline waters below, in which flourished sponges, corals, clams, mussels, whelks, crabs, sea cucumbers, starfish, octopus and fish. Each summer, the ecological boom drew narwhals, walruses, seals and beluga whales, while sea birds made a 25,000 mile round trip from Antarctica. The tundra came out in a spotty cover of lichens, mosses, dwarf willows, scurvygrass, avens, saxifrage, lousewort, sedges, rushes, sphagnum moss, dwarf birch, sorrel, crowberry and buttercups. Arctic hares, lemmings and rock ptarmigans ate the vegetation. Snowy owls and Arctic foxes ate the lemmings and voles. Caribou ate lichens and birds' eggs. Musk oxen ate wild grasses and willows. And wolves ate caribou and musk oxen. In the taiga, the dry hardwood forests burned in the summer heat – an essential part of their survival – but they also flowered and fructified with enough food to support foxes, wolverines, lynx, moose, snowshoe hares, owls, goshawks, chickadees, tits and woodpeckers.

Perhaps, in total, three thousand mammal species lived north of the Arctic Circle, though only a few dozen lived in the permanently frozen Arctic. Those mammals which did evolve under those conditions, specifically to cope with ice, permafrost, scant nutrition and sub-zero temperatures, tended to be the most charismatic. The beluga whale, all white to camouflage itself against predators, without dorsal fins to swim freely under ice cover, and with a special cranial unit for echolocation to navigate under the ice. The ringed seal, which evolved during the last glaciation, with exceptional underwater breathing, and strong front flipper claws to cut breathing holes through the ice and excavate caves in snow drifts for birthing lairs and protection from polar bears. Polar bears, the glamorous icon of ice-dependent animals, diverged from grizzlies 600,000 years ago, when ice covered much more of the planet than it does today. A bear population separated itself and drifted off into the icy north, where – in splendid reproductive isolation, and without competitors for energy-rich seal meat – it evolved under intense selective pressures.

In place of the pie face and broad nose of the grizzly, there emerged the narrower face (for catching prey in burrows and breathing holes during the Arctic spring) and longer snout (for warming Arctic air in the nasal passage). In place of highly visible brown fur, white fur became the norm. Feet grew larger and became overgrown with hair. Heat-losing extremities like ears and tails shrank, ice-gathering whiskers disappeared. They developed a layer of outer guard hair to prevent hypothermia. Lipovorously larding on bulk, while remaining agile runners and swimmers, their bodies grew wedge-shaped.[4] Though faithful to ancestral feeding and birthing grounds in the polar desert, they ranged wide in their hunting, and even descended to graze and socialize in the high Arctic tundra during the summer months. When hominids appeared, they were unafraid. Indeed, they were the only animal to hunt humans. These apex predators moved, John Muir writes in *The Cruise of the Corwin*, 'as if the country had belonged to them always'.[5]

The old Arctic is dying, hence its perturbations. According to the National Ocean and Atmospheric Administration, the warming of the Arctic is introducing a fundamentally new climate system.[6] Ancient ice shelves are cracking up, collapsing and calving into bergs. The permafrost melt is now at a level it was once predicted to reach in seventy years' time. The most pessimistic predictions of an 'Arctic death spiral', as the geographer Mark Serreze calls it,[7] now look sober. The bears, with less first-year ice from which to hunt lipid-rich seal, walrus and whale meat, are turning to snow goose eggs and caribou. The selective pressures of climate change on bears and walruses, ivory gulls and kittiwakes, seals and beluga, are already hitting hard and fast. The Inuit term for the ecological loss, the depression, the environmental melancholia, the solastalgia induced by the loss of ice, of habitat, of home, is 'uggianaqtug'.[8] It has connotations of a friend behaving oddly, both frightening and sad. It is from this experience that recent research into climate mourning emerged. Yet not all mourning is the same. We all have something to lose in the Arctic, whether we know it or not. Yet there is a form of environmental lugubriousness, a form of sighing nostalgia which one might associate with Attenborough and certain English ecologists and writers, whose ecological consciousness is suffused with the polar sublime.

The polar sublime is far too entangled with all that which conspired to kill the Arctic. In retrospect, despite its emergence from a Romantic sensibility, it was a sort of Fata Morgana, a colonial, anthropocentric lure and, later, an emission of fossil capital. It needed an essentialized, eternal polar landscape in which character could be tested, and which would never stop yielding its abundant supply of fish, whalebone and whale oil. It was not merely arrogantly dismissive of the northern peoples – 'the most wretched of Beings', Keats recounted,[9] describing the second-hand account of an expedition led by Captain James Ross – but profoundly optimistic in its assumption that the ends of the earth, with their exacting trials and holy sanctuaries, must always have been there, and would always be there. However deadly and dangerous these polar journeys may have been – 'never again', said both Apsley Cherry-Garrard and Richard Byrd of their Antarctic voyages[10] – however unsettling and defamiliarizing they were, the experience was carefully curated so as to offer reassurance as to the indestructibility, the godliness, of an underwhelming race of hominids. And specifically, of course – even as explorers devolved into the most interestingly non-contemporaneous ideas and fantasies, from clairvoyance to spiritualism – as to the manly, intrepid scientific-ity and intellectuality of the white bourgeoisie. The homeland of such polar non-contemporaneity in the twentieth century was New Swabia.

And so this is an obituary, not for the Arctic, some of which may be salvageable, but for an idea of the north. Ultima Thule, septentrion, the boreal. The polar imaginary with its god-appointed 'vast alabaster whiteness', its 'white Mars' landscape, its empty 'desolation', its church-like 'silence', its mystical solitude, its self-defining colonial encounter with the savage and unmastered. For a new polar sublime.

DISASTER
AND DENIALISM

18 FEBRUARY 2021

Climate disaster intensifies denialism. In Oregon, the wildfires were blamed on 'antifa'. In Texas, the storm-induced power outages are blamed on 'environmentalists'. This is a rule of the capitalocene, and we should expect more of it.[1] If we look closely at what has happened in Texas, we can see how this works.

The wandering of the polar vortex brings an ice storm to Texas. Temperatures that usually average about 17 degrees Celsius during the day fall to −5. The majority (about 67 per cent) of the electricity is supplied by coal and natural gas. A minority (about 21 per cent) is supplied by wind and solar.[2] The pipes and stations are not insulated against the cold. The pipes freeze, stations are knocked out of action. About half of the natural gas capacity is wiped out,[3] as are most wind turbines. Most of the energy loss is due to the disruption of thermal sources like coal and gas. To avoid a complete shutdown, the local energy authority, ERCOT, asks suppliers to implement controlled outages. Texas homes are not designed to stay warm in winter. So when millions go without any electricity, people start freezing to death in their homes. Some people, in a desperate effort to ward off hypothermia, run cars and generators indoors – and die from carbon monoxide poisoning as a result. Water provision is also compromised: millions of Texans are under a 'boil water' notice, because burst water pipes means they can't rely on the cleanliness of their drinking water.[4]

There is no such thing as a 'natural disaster'. From Katrina to the plague, we have seen that the 'natural' part of the disaster is increasingly an industrial by-product. So it is in Texas. The heating of the Arctic means that the polar vortex will wander south more often, and bring more extreme weather events with it. This has already been happening for a while.[5] But even the snowstorms are not inherently disastrous. Freezing weather is spreading through the Midwest and out to the East Coast. In Texas, snowstorms are a disaster. This is because of socially determined vulnerabilities. In most of Texas, despite prior warnings and past experience, the energy system is not 'winterized'.[6] There is little incentive, in a privatized, streamlined-for-profit energy system, to take such precautions, and little risk of energy firms being threatened with regulation. Moreover, the Texas grid is more vulnerable to such abrupt shortages because the state is an 'energy island' relying entirely on its own grid due to its determination not to be subject to federal regulations. Thus, the disaster breaks.

Next comes the disaster management. It's going badly. Why? Texas has its own emergency management department (TEDM). This was launched after Katrina, when local authorities concluded that in the event of a disaster FEMA would do nothing for them. Among the guiding assumptions of TEDM's basic plan is that federal assistance will not be available and the government resources will not be adequate to meet disaster needs.[7] Like FEMA, it relies on outsourced and voluntary efforts. This is partly due to state failure. For example, citizens' corps are relied on to make up for shortfalls in firefighting and healthcare. Similarly, charities like the Salvation Army are relied on to manage shelters and feed people. Government resources are available, but the assumption is that local city and county authorities must exhaust their own resources, including whatever aid they have available, before asking the state for help. So TEDM sees its remit being largely to coordinate a series of fragmented emergency responses, both so that Texas doesn't end up as helpless as Louisiana in 2005, and so that it doesn't rely on federal intervention.

This sort of patchwork emergency response amounts, in practice, to almost nothing. That is why people are dying. People are having to melt snow on gas cookers in order to have water to wash with. They are having to phone round hotels to see if they can find affordable accommodation before their pipes burst or the ceiling caves in. They are having to improvise individual responses, as reported in the *Guardian*, in the absence of a serious state response.[8] The main policy response from the state has been to ask stranded drivers and those freezing in their homes to make their way to 'warming centres' run by businesses or the Salvation Army. No proper food provision, no alternative accommodation and very little information. That's bad enough, but many of those warming centres lost power in the outages. And faced with a complete lack of information from ERCOT or TDEM about how to handle the rolling outages, or how long they will even last, the only answer that the Austin mayor had was rationing. In the deep freeze, he asked those who do have power to live as if they didn't and keep the heating turned way down. The situation would be even worse, had the governor of Texas not secured an Emergency Declaration from Biden.[9] I'm not sure that he would have obtained one from Trump, but that declaration means that Texas can now rely on 75 per cent federal funding, and that FEMA is sending basic supplies like blankets, meals and generators. Now, this is just a sticking plaster, and far from adequate, but it is telling that a system designed to bypass reliance on FEMA would have collapsed without it.

How does denialism work in this situation? It's been widely remarked that Texas Republicans have come out fighting, by blaming renewables and the 'Green New Deal' for their situation. Texas governor Greg Abbott blames the energy shutdown on renewables.[10] Former Governor Rick Perry claims that 'if this Green New Deal goes forward the way that the Biden administration appears to want it to, then we'll have more events like we've had in Texas all across the country'.[11] There is no point in fact-checking these claims. Everyone knows that Texas's problem does not come from its renewables, and that Biden has no plans for a Green New Deal. That isn't the point. The point can be found in the less noted argument that Texans want hardship, welcome hardship, if it means keeping the federal government off their backs. Thus, again, Rick Perry, who claimed in an interview with the Republican Congressional leader Kevin McCarthy's website that 'Texans would be without electricity for longer than three days to keep the federal government out of their business'.[12] This is structurally identical to the claims, early in the Covid crisis, that grandmothers and grandfathers would be happy to suffocate to death in overcrowded hospitals if it meant their children could continue to enjoy capitalist freedom.[13]

This is not a *factual observation*. It is an *interpellation*, a call to arms. In a serious climate mitigation crisis, it summons a transversal alliance of the right in opposition to any politically correct plans for human survival. To say that some are happy to die for capitalist freedom, is to say that they are disposable. The hard morality of disposability has long been built into fossil capital, but now it acts as a rallying point. They're saying the quiet part out loud because they need a mobilized political constituency that is ready, even morally energized, for quite a lot of death.

If you want an exemplary case of the implicit authoritarian and social Darwinist values of petromodernity being made explicit, consider the mayor of Colorado City, Texas, who recently resigned. He was frustrated, not by his inability to do anything, but by the lazy, handout-seeking bums who didn't understand that it's 'sink or swim', and that 'only the strong will survive'.[14] It was pleasantly surprising that he had to resign, as that hasn't been the rule in recent years. Because, after all, the mayor didn't say anything that hasn't been affirmed countless times by successful politicians, presidents, newspapers, think-tanks and top-flight periodicals. His mistake, perhaps, was that he said it in a context where it couldn't plausibly be read as merely a statement about an underclass or a minority. Had there been some 'looters', or some social disorder drawing the attention of militias, he might have enjoyed greater popularity. However, in this case he was clearly talking to the *majority* of his constituents.

The molecules of what Cara New Daggett calls 'fossil fascism' are bonding into complex compounds.[15] And what one should look out for is not the obvious material investments that people might have in prolonging fossil capital. Certainly, people want to cling on to what they have, materially and symbolically. However, for the interpellations above to work, they have to be offered something more than that. If they want a section of poor and working people to start saying, 'Yes, I will die for this', fossil fascism has to offer them something that is *worth more than survival*. And once we're in that terrain, we're always talking about the satisfactions of revenge, destroying a neighbour, enjoying omnipotence over another life, even if vicariously.

NOTHING
BESIDE REMAINS

31 MARCH 2021

I.

After the lockdowns, the struggle for an ecological reconstruction has to revive. It's unclear what the appropriate political format for that struggle will be. Until it was knocked out of action by the pandemic, XR's style of anti-political street theatre looked likely to dominate. Its dormant infrastructure can probably be reanimated, but there will be opportunity costs if it just does the same old thing.

XR is exemplary of what Christopher Bickerton and Carlo Invernizzi Accetti define as 'technopopulism'.[1] On the one hand, it summons a virtuous people to dislodge or goad into action a decrepit political leadership. On the other hand, it appeals to a 'truth' that is supposedly beyond left and right. This truth comes in two essential parts. The first is that the physical danger to humanity is so pressing that civil society sectors can only retain their attachment to the fossil economy by remaining in the dark, or in denial. The second is that the findings of a major study of civil resistance by Erica Chenoweth and Maria Stephan purportedly show that the threshold of political change is met if you can non-violently mobilize 3.5 per cent of the population.[2] Thus: enlighten the people, mobilize a critical mass, and get change. In its best light, this tendentious reasoning allowed XR to put tens of thousands of people on the streets and kept open channels of public sympathy that probably restrained attacks from the police. That's not negligible, given that the police regard XR as a 'terrorist' offensive against the state.[3] Yet this had probably reached its limits before the pandemic. And after what we've just been through, it is simply no longer plausible to hope that the mere apprehension of a physical threat is sufficient to make people think rationally about survival.

More troublingly, if the organized expressions of ecological politics don't confront the material foundations of the energy system, and thus of the energy revolution that we require, they will find their message co-opted and neutralized. At this year's COP26 in Glasgow, for instance, the UK government will claim that it can manage the shift to a decarbonized, energy efficient economy with minimal disruption to capitalism. That is, it will claim to be able to reach 'net zero' emissions within its territory by combining electrification of transport, the expansion of renewables and nuclear power, the improved management of natural resources and the deployment of carbon capture and storage. As the public accounts committee has made clear, there is as yet no meaningful plan to achieve this goal. Nonetheless, the rhetoric of acknowledgement, of green investment and green jobs will be part of the government's pitch. It can claim, if no one probes too far beneath the surface of rhetoric, to be 'telling the truth' and even 'acting now'.

It might even get to the stage of offering some sort of citizens' plebiscite, thus sort of fulfilling the third of XR's demands.

II.

Realistically, any plans for 'net zero' would be plans for 'absolute zero'. There is some scope for reducing emissions in the natural environment (which comprise about 12 per cent of total UK 'territorial' emissions[4]), but it's not likely to become a carbon sink in the next thirty years.

Carbon capture and storage (CCS) technology is essential. However, the UK's plans to develop CCS using subsidized private enterprise, as in Longannet and Kingsnorth, have never borne fruit.[5] The policy of getting CCS deployable by 2020 and retrofitting all coal installations by 2025 isn't close to being met. Much of the 'net zero' target will be met with the use of nuclear power, which is a costly, short-termist option imbricated with military-industrial power and relying on tendentious industry claims as to its low carbon-footprint. As for the goal of controlling emissions only in UK territory, this is just the auto-infantilizing colonial idea that someone else can be made to clean up our shit. The UK offshores many of its emissions as a matter of industrial policy, so that a focus on territorial emissions obscures its true metabolic profile.

As a major 2019 report by the UK FIRES consortium suggests, a 'net zero' emissions UK would only have access to about 60 per cent of the energy that it currently uses by 2050.[6] This isn't as bad as it sounds, because there is a great deal of room for efficiency savings in cars, heating and new-build homes. Electricity is also more energy efficient than fossil fuels. However, even the most optimistic scenario for renewables development will not solve the problem of shipping, flying, construction and land use. Flying will have to stop until an electrified or sustainably biofuelled airplane has been invented and produced at scale. But every single UK airport has plans to expand, and these will be backed by the Conservatives, quite a lot of Labour MPs, and even the leftier end of the trade union bureaucracy. Merchant shipping is essential to the world economy as it exists and would need to be electrified: but the technology doesn't yet exist for that. Construction would have to find alternatives to cement because of the emissions created during the chemical reaction of calcination. Beef and lamb, the major source of emissions on UK farms, would have to be phased out in favour of a much more vegetarian diet. These plans require not only an industrial strategy of a kind that the UK hasn't had for decades to manage a jarring rupture in production relations, but a radical recalibration of social priorities and ways of living.

The energy revolution required here is profoundly unusual in human history, in that it involves a shift from more concentrated to less concentrated sources of energy, and from more to less energy consumption. It also necessitates that, when we do have the technology to sequester carbon from the atmosphere, we do not take that as a reason to emit more; which is unlikely, for the same reasons that the Jevons paradox shows us that more energy efficiency doesn't mean less energy consumption.[7] Assuming that we don't achieve a socialist revolution in the next thirty years, let alone the next decade, this means that we need to find a way to make capitalism energy efficient. Which is a contradiction in terms.

III.

The specificities of what the ecological economist Jean-Claude Debeir defines as the capitalist 'energy system' are such as to make energy efficiency impossible.[8] Any energy system combines the ecological and technological characteristics of chains of energy with the social structures which appropriate and manage both energy sources and energy converters. Central to every mode of production's relationship to energy is how it appropriates and manages human converters, for that is how sufficient material surplus is generated to sustain a ruling class. The management of human converters, through education, training, deployment and renewal, could be called social energy. Like physical energy, its efficient deployment depends on its being assiduously concentrated and renewed, which is itself an expenditure of energy. This combination of labour with natural energy is what the autonomist philosopher George Caffentzis called the 'work/energy regime'.[9]

Pharaonic Egypt centralized village-bound human muscle power and harnessed it to hydraulic power through the technology of a centralized religious, tributary state. In feudal Europe, the lords competitively harnessed human and animal muscle power to a hydraulic energy chain to extract tribute from the growing and grinding of grain, and the manufacturing of tools and textiles. The capitalist mode of production extracts not tribute, but surplus labour. It does so by organizing competing labour forces to leverage and convert physical energy into commodities. The surplus labour embodied in them can be converted into profit only through success in the sphere of commodity circulation. This process, being driven by the competitive accumulation of capital, with its creative-destructive cycles, lends itself to accelerated entropy of both physical and social energy. The majority of commodities will not re-enter the loop as renewables: if they did, it would reduce the opportunities for profitable investment, hence such phenomena as 'planned obsolescence'. The routine upheavals in production, moreover, as investment resources are directed away from unprofitable areas towards more promising prospects, entropically scatter social energy in the form of skills, training and experience.

Capitalism was hardly the first mode of production to discover and exploit fossil fuels. Indeed, the first decades of the Industrial Revolution were driven by hydraulics. However, as the geographer Matthew Huber argues, the rediscovery of highly concentrated, fossilized solar energy was strategically central to the real subsumption of labour power by capital.[10] The fossil-powered mechanization of tools allowed capital to expel workers from control over the production process, while increasing the extraction of relative surplus value through better labour-saving devices. It also permitted capital to geographically concentrate production, while extending the spatial orbit of commodities. With fossil energy, it was able to decisively break through the previous biological and ecological limits of work. This, however, simply made it easier to waste vast amounts of energy.

This does not mean that capitalism used more than a fraction of the energy that is available. The earth is an open energy system whose resources capitalism uses poorly. The economic addiction to fossil fuels, and the concentrated political and monopoly power that congeals around their extraction, has probably impeded the efficient concentration and renewing of resources like solar energy and tidal power. Less than 0.05 per cent of solar energy is transformed by photosynthesis into the basis of all living matter, and less than 0.02 per cent of it is managed by humans, according to Bent Sørensen.[11] This is because the gift of solar energy is diffuse, and difficult to concentrate. Nonetheless, the technologies for concentrating solar energy are becoming cheaper to make. Between 2000 and 2017, the cost of producing a solar PV panel fell 94 per cent, to $0.29 per watt.[12] The main operating cost for solar energy, as with wind power, will come from siting. In principle, tidal power could supply most of our energy needs if it could be adequately concentrated and stored. Like most hydro-power projects, this would require the commitment of the national state, as in France and South Korea, which collectively make up 90 per cent of the earth's tidal power. Nonetheless, pending investment and site location, researchers at Oregon State University have suggested that just 0.2 per cent of the ocean's untapped wave energy could supply the world's current energy demand.[13]

IV.

In principle, we humans can live together a lot more energy efficiently, and justly, while recognizing our different values. Surveys and time use studies cited by FIRES suggest that the activities and foods we most enjoy don't use a lot of energy. We could make some trade-offs, for example giving up the energy-rich consolations through which capitalism secures popular consent, in exchange for improved quality of life and health, and a better communal life. This would have to involve a conscious, collective effort to redefine what is enjoyable, what is delicious, what is abundant, what constitutes a good life. But it would not necessarily involve an austere life, because, in principle, we could sustainably harness a lot more energy for the benefit of many more people than we do.

However, the contradiction in terms that is energy efficient capitalism seems to be both an impossibility and a necessary staging post on the way to such a future. It wouldn't be sufficient, either, for capitalism to stop growing for it to become energy efficient (as in a steady-state economy). The investment decisions would have to be subordinated to political control, pushing Keynes to his utmost radical limit. And, of course, if this wasn't won in a democratic way, it would simply lend itself to authoritarian state capitalism in which the goal of energy efficiency would be subordinated to the self-realization of a new statist elite. Further, the end of growth would bring a major confrontation to a head. Growth is what makes capitalist inequality notionally compatible with declarations of the moral equality of all human beings. Without growth, the principle of equality either has to be honoured with some redistribution, or violently dispensed with as the rulers seek to defend their monopoly on wealth and its production.

By arguing like this, I could be at risk of reproducing XR-style apocalyptic moralism. Except that, instead of faulting a political leadership, I would be blaming everything on capitalism. Of course we can do better than capitalism. But the point about capitalism is that it's something we all, in different ways and to different degrees, are involved in. We reproduce it, we are implicated, we have attachments to it, it has a share in our dreams. What we call 'petromodernity' is just the name for one genre of capitalist dreaming. This is a strategic factor in struggle.

The danger of apolitical technopopulism is that, by relying on 'end of days' apprehension to achieve its mobilizing effects, it doesn't prepare any of its target audiences for the complex negotiations and struggles ahead. By prescribing a citizens' assembly without spelling out what its supporters should fight for in such assemblies, it leaves such techniques wide open to rightist, denihilist or ecofascist appropriation. By only training its cadres in the most basic of protest techniques and the most shallow, morally frivolous forms of social movement theory, it leaves them disempowered in the face of the powerful and sometimes subtle offensives they will face from all sides. So, if we need to do better than capitalism, and even if we need to better than the current UK government, then we also need to try out new organizing bids, seek out new leadership, speculate on new tactics, and above all raise the level of political understanding.

WHAT THE
EYE LIKES

7 APRIL 2021

Capital has an apparatus of guilt that is related to debt. The innovation by fossil capital of the idea of an individual 'carbon footprint', another way of turning collective responsibility into personal debt, was a far-sighted investment in guilt.[1] This is a strategy with deep roots in fossil capital. And it is now wired into what Kathryn Tanner calls the 'new spirit of capitalism'.[2] The disciplinary effects of financialization on the whole system come at us in the form of a seemingly chaotic buffering of unaccountable forces, and yet we live their effects with a completely individualized burden of guilt for them. This is an example of the way, paradoxically, we feel most guilty when we suspect we have no say in what happens to us.

The related modern concept of 'public apathy' has been imported into environmentalism from the annals of marketing. It was originally an attempt by marketers to label the non-responsiveness of consumers and target audiences to a specific campaign. We can see how this ideology works to incapacitate effective action when even evidence of clear majorities favouring necessary measures is packaged as evidence of public indifference.[3] And we can see the other side of the double bind when resistance is classified as 'extremism'.[4] Do nothing, you're guilty; do something, you're guilty. Guilty, guilty, guilty. The only people who aren't in the least bit guilty are the owners and managers of the hundred corporations responsible for 71 per cent of global carbon emissions.

What we call 'apathy', Lertzman argues in *Environmental Melancholia*, is no such thing. In her fieldwork, she finds not dead-eyed doziness, but an anguished compound of stalled mourning, ambivalence, anxiety, powerlessness and the guilt that follows from powerlessness. It might be even better to speak of a kind of 'anticipatory mourning', wherein, perceiving the damage to an object of our love (what the Gwich'in call the 'wounded sacred'), we start to withdraw affection from it, before it finally dies. An example of anticipatory mourning appears in Freud's short essay 'On Transience' (1915). He describes a walk through a 'blossoming summer landscape' with a poet friend, presumed to be Rainer Maria Rilke. The poet expresses admiration for the beauty, but dared not be moved by it because it was bound to fade and die. Within a year, the ecological cataclysm of the First World War erupted and 'robbed the world of its beauties'. Yet that did not devalue the beauty that had existed. To respond to transience by devaluing what will surely pass, to speak of one's enviro-somatic envelope in the past tense, as something that once held one securely, is in Freud's terms a 'revolt' against mourning.

And yet it comes. Tim Gordon, a marine biologist, describes in a BBC news report on Radio One how 'occasionally, for no particularly good reason, it'll strike – you just float into the middle of the water, look around you and think: "Wow, it's all dying." There's been times that you cry into your mask because you look around and realise how tragic it is.'[5]

II.

Walking such paths, you might walk up strange pasts. This in the hunter's sense of 'walking up' – meaning *to flush out, to disturb what is concealed.*

ROBERT MACFARLANE, *Holloway*

If climate activism frequently takes an apocalyptic tone, stuffed with theological expectancy, it is because, as Catherine Keller argues, 'theology […] articulates what unconditionally matters'.[6] The planet is what unconditionally matters, since it is the undercommon ground of all life. We can't simply add 'ecology' to a list of issues concerning the left, because it is the unconditional condition for everything else.

It's all dying, and somehow it isn't enough to point out that 'consumers' are not to blame. To what imaginary tribunal are we pleading innocence, anyway? Besides, we're all implicated, because the infrastructure of life depends on fossil capital. And even throwing ourselves into activism can leave us feeling even more helpless, even more guilty, because of how massed the forces are against change. 'Doing something', even if it doesn't necessarily help, becomes a way to master that feeling.

Anxiety is a threat-response. It needn't be a present threat, or even a real threat, as long as we believe in the idea of the threat. In fact, anxiety in the properly psychoanalytic sense is a response to an unknown, enigmatic danger. What else could we call the perma-threat of wild weather events, new epidemics, new economic crashes, all both predictable and unpredict-able? How much more disturbing a power might such possibilities have if they merged with the signal of an inner, psychic danger, which followers of Melanie Klein call the 'death instinct' and followers of Lacan call the 'real'? And what if the threat doesn't ever go away? As Anouchka Grose tells us, in her *Guide to Eco-Anxiety*: the hypothalamus keeps ordering more cortisol, more adrenaline. The heart beats faster, the lungs open up, immune and reproductive systems are suppressed. Anxiety is supposed to be transient, not permanent. The longer it goes on, the higher the blood pressure, the more immune problems arise, the lower the bone density, the higher the risk of strokes, the greater the exhaustion, the lower the fertility rate. When Dominic Pettman wonders, in *Peak Libido*, how it is that celibacy rates are soaring and sperm counts are down, pervasive anxiety may be part of the answer.[7]

If anxiety is derided as overthinking, the prescribed cure of exercise is supposed to neutralize thought. A cheap, depoliticized way to get people to complete their fight-or-flight responses in quiet and piss off. It doesn't work. Run off some stress, get an endorphin cookie, and you'll still face the headlines about wildfires and mass extinction. It's all still dying. However, *thinking* is an ambulatory experience. The muscles, heart and lungs partake of thought, conscious and unconscious. 'Embodied cognition' is not the half of it. To go walking down coastal paths and up glacial peaks is to walk up strange pasts. This is what Nietzsche intuited when he said, in *Ecce Homo*, 'do not believe any idea that was not born in the open air and of free movement'. This is what Maurice Merleau-Ponty adverted to when he suggested we 'do a psychoanalysis of nature'.

III.

We have to cultivate sensibilities as much as ideas and strategies. The ambulatory doctrine of thought is a counterpoint to the anti-naturalism, indeed socio-centrism, of radical social theory. Theory that resists biochemistry, oceanography and palaeontology tends to undercut the necessary sensibilities. It tends to down-grade 'nature' and 'wilderness' as so much romantic, völkisch twaddle (of which, indeed, there are libraries full). It perhaps tends, in this way, to resume the cold circuit of idealism. To 'do a psychoanalysis of nature' is to work through the chiasmic relationship between 'spirit' and 'nature'.

Bollas's term for strange pasts is 'the shadow of the object'. 'It is usually on the occasion of the aesthetic moment,' he writes, 'that an individual feels a deep subjective rapport with an object (a painting, a poem, an aria or symphony, or a natural landscape) and experiences an uncanny fusion with the object, an event that re-evokes an ego state that prevailed during early psychic life.'[8]

The aesthetic moment is not what we might have supposed. Marion Milner, in learning how to paint, had to give up on the inhibiting principles of beauty, perspective and realism.[9] She would sit in front of a landscape, and produce technically proficient, realistic paintings. But they left her feeling that they were hardly worth the bother. They were lifeless. The common-sense foundations of sensory experience denied her creativity. Then she discovered that 'the eye should find out what it liked'. And when she did seek out what the eye liked – a feature of the sea wall, the way outlines appeared to melt away – she found that somehow they expressed moods, feelings deriving as much 'from the sense of touch and muscular movement' as from sight. In Santayana, she reads that 'waking life is a dream controlled' and that 'the gods sometimes appear' in nature. As Milner discovered, much of what began to appear in her art, composed in moments of apparent serenity, was 'fire and tempest'. It was what Merleau-Ponty would have called the 'barbaric principle'. Or that signal of internal danger that one might call the death drive, or the Lacanian 'Real'.

In Bollasian terms, the natural environment is a version of the 'transformational object' *par excellence*. The transformational object is the earliest emotional tie in an infant's life, the first caregiver, usually the mother. What makes this object different from others is that the infant at first experiences the mother not as a separate object, but as its *environment*. It finds its moods regulated by changes in the environment, by nurture and play, by giving and withholding, by presence and absence. And because this relationship is pre-linguistic and pre-mirror phase, the object will be recalled existentially, not representationally. In adult life, we can strive to rediscover this object within a mythic or religious structure, or by buying a commodity which is advertised as a solution to our difficult feelings (feelings which the advert will have deliberately incited or exacerbated). Wherever we find this object, as we may find it lying about in nature, it produces an uncanny feeling of recognition.

In my own experience of walking up strange paths, I try to memorialize whatever has produced a stir. As I can't eat the scene, I take photographs, and I make lists of names and descriptions. Wild, windswept Cornish farms, lines of frosted wild cherry and alder, walls of slaty mudstone rising like a great cliff, lemon blooms of gorse, the plump grape-coloured berries of blackthorn, the marshmallow-pink Cornish daises, the spiky olive fronds of *Astelia* and *Furcraea*, slates of watery blue sky in muddy fields, the curious symmetry of equally desolated, fossil-black columns stuck with seaweed, a booming Atlantic thronging with the compound music of ruddy turnstones, skuas, puffins and storm petrels.

It's all dying. Visit it, as you would a dying patient.

SCALA NATURAE

31 MAY 2021

I rejoice that horses and steers have to be broken before they can be made the slaves of men, and that men themselves have some wild oats still left to sow before they become submissive members of society.

HENRY DAVID THOREAU, *Walking*

I

Every time I hear of an escape from the zoo, I burst out laughing and punch the air. Just over a month ago, two dozen monkeys escaped from a zoo in Berlin, and the zookeepers couldn't understand how. Such escapes are far more common than you would think, and often display considerable resourcefulness. In San Diego Zoo, an orangutan was born, and they decided to name him Ken Allen. From infancy, Ken was unscrewing nuts, climbing humans, finding every expedient he could to escape. When they brought in Vicki, a female mate, to distract him, she unbolted a door and escaped. One evening a few years back, in the National Aquarium of New Zealand, an octopus named Inky climbed out through a small gap in his tank, ran across the floor, and then climbed down a 164-foot drainpipe into Hawke's Bay.

The laughter only stops when I remember the fate of the golden lion tamarins. They had been driven to near-extinction in the forests of Brazil, with only a few dozen preserved in zoos across Europe and North America. In 1984, conservationists started to reintroduce some of their captive-born golden lion tamarins into Poço das Antas Biological Reserve in southeastern Brazil. The experiment was, at first, a terrible failure. They died of snake bites, bee stings and starvation, sometimes within days of release. If the animals were pure mechanical creatures of instinct, without culture or society, they would presumably have adapted straightforwardly to the conditions that they had evolved in. But they were and are cultural beings, and had learned to live in the peculiar environments of zoos. For the first decade, every reintroduction of the animals met with the same grim results. Only when their release was carefully staged, and care slowly withdrawn to enable them to find their way in the new environment, did the reintroduction work.[1]

All of which is to say, liberation is not as simple as escape. These animals had been drawn into a complex social relationship with their human handlers and needed to be allowed to negotiate their way out of it.

II.

In the twentieth century, the scientific mainstream in the imperialist nations considered it controversial to impute to non-human animals (henceforth 'animals') ostensibly 'human' qualities like imagination, desire, fore-planning, playfulness. This was deemed 'anthropomorphic' by behaviourist-inflected ethologists and biologists.

That little term, 'anthropomorphic', expediently supplied its own proof. To invoke the term wherever certain types of intelligence were applied to animals was to surreptitiously classify those intelligences as specifically human. Animal consciousness must, from this perspective, be considered a 'black box'; all we know about is stimulus and response; all we can infer is instinctual machinery. This is a version of the old Cartesian picture of human exceptionalism, or the Aristotelian *scala naturae*, the great chain of being. Lately, under the burden of accumulating scientific discovery, and with the growing awareness of our ecological dependencies, that sensibility has broken down.

The work of biologists, philosophers, ecologists and popularizers like Donald R. Griffin, E.O. Wilson, Marc Bekoff, Eva Meijer, Carl Safina, Peter Godfrey-Smith, Peter Wohlleben, Frans de Waal and others has brought to human consciousness the complex varieties of animal consciousness and emotional life.[2] They have told us that certain animals use words (rather than merely imitating them), that fruitbats and dolphins in the wild have names for one another, that sperm whales communicate their identity to one another with rapid bursts of clicks, known as 'codas', that make the water vibrate for several cubic miles. 'So powerful and penetrating are their sonar clicks,' Safina writes in *Becoming Wild*, 'that sperm whales can likely see what many things look like inside, as if X-raying them.' They have told us that prairie dogs have a complex, open, language-like communication system that allows them to describe any humans nearby down to their size, clothes, hair colour, and any objects they're carrying. They have shown that animal communities have moral codes in which those who play foul are shunned, and that the specificities of the code varies between communities, not just species. Even what we call 'instinct' cannot be mechanical. The rabbit may be 'programmed' for flight when a predator is near, but if her line of flight displayed no creativity, no contextual awareness, she would be eaten.

They have told us that chimpanzees have fashion trends. Primatologists from the Max Planck Institute for Psycholinguistics in Nijmegen observed this when Julie the chimpanzee started to wear a blade of grass in her ear, and other chimpanzees began to follow her. They have told us that monkeys and birds have the same mirror neurons that humans have, cells that appear to be involved in empathy and self-consciousness. They have found that whales have the same spindle cells that humans have, the cells that allow us to love and suffer emotionally, only they have many more of them and have had them for longer. That elephants and corvids hold funerals, and that whales and red deer grieve. That dolphins play games with objects found in the ocean, and that they spend more time playing than hunting. That cephalopods experience emotional, not just physical, pain. That cetaceans display rapid eye movement when they sleep, suggestive of dreaming.

Whales have been dreaming, that is, for almost fifty million years before humans arrived. This is the vast 'whale brain' which Heathcote William rhapsodizes about in *Whale Nation*. We are speaking not merely of cognition, but of consciousness, of beings capable of love, play and mourning. Not merely of mechanical chirruping, gesturing, clicking, calling, scent-emission and dancing, but of complex and often generative systems verging on what we call language. The animal, *contra* Martin Heidegger, is not 'poor in the world'.[3]

III.

In *Are We Smart Enough to Know How Smart Animals Are?*, de Waal wonders aloud how it was that science chained itself to inhibiting, dangerous assumptions about animals for so much of the twentieth century: assumptions that would have been utterly baffling to a Victorian rationalist like Darwin. This is a very good question. After all, we have as much evidence for animal cognition and emotion as we do for that of our fellow humans. It's a curious form of species solipsism to believe that we are unable to infer intelligence, desire and dreaming in animals based on how they communicate and behave.

The answer to de Waal's question may have something to do with the problem of mass extinction, to which we were first alerted by Richard Leakey and Roger Lewin a quarter of a century ago in *The Sixth Extinction*. Perhaps the exclusion of animals from emotional, cultural and linguistic life, and therefore the sphere of moral consideration, had something to do with the atmosphere of Cold War triumphalism in which the work of the natural sciences was conducted. What Victorian science did contribute to the mid-century temperament was a cosmic productivism. The new energy science responded to the central problem of industrial capitalism, which was how to sustainably and profitably (and if not both, profitably) convert bodies into machines: *The Human Motor*, as the historian Anson Rabinbach puts it. Yet the implications of energetics made the whole universe a productivist furnace. Everything, not just animal life, was a machine regulated by flows and transformations of energy, by conservation and entropy. Life itself was machinic.

As Troy Vettese argues in his *Salvage* essay 'A Marxist Theory of Extinction',[4] this view of animals as machines is an axiom of capital, rendered fully explicit in neo-liberal writing on nature and the commons. The 'real subsumption' of biological life by capital converts it into a machine, to be manipulated and set to pace and productive rhythm like any other technology.

In Foucault's terms, we could characterize this as the biopolitical axiom of 'making live' and 'letting die'. Thus, as in the practice of salmon farming,[5] what Kenneth Fish calls *Living Factories*. Those species which are not subsumed go extinct.

Contemporary culture is probably no more innately cruel to animals than the cultures of previous class societies. The Roman empire, Ashley Dawson tells us in *Extinction: A Radical History*,[6] was catastrophic for animal life. Its rich metropolitan life was sustained by agricultural extractivism in the colonies, necessitating deforestation on a grand scale. Its haute cuisine of thrushes, wild boars and flamingo entailed an international trade in wildlife as luxury. Its Colosseum entertainments entailed the massive slaughter of lions, leopards, bears, elephants, rhinos and hippos: 9,000 animals in the three-month period after Titus dedicated the Colosseum.

Yet it is capitalism, not ancient slavery and empires, that brings us to the point of endangering one million species. It is capitalism that has brought about the biological annihilation of insect species that threatens global food production. It is specifically capitalist fishing that is killing the right whale, the hawksbill sea turtle and the vaquita porpoise, most often because they are entangled in dumped fishing gear, the biggest source of plastic in the oceans. In *The Tragedy of the Commodity*, Stefano B. Longo, Rebecca Clausen and Brett Clark outline the history of fishing and modern aquaculture, noting that while human societies have been dependent on marine resources since the end of the Pleistocene, this did not assume the shape and proportion of a physical threat to species, leading to the ubiquitous implosion of fisheries, until the Industrial Revolution. The take-off since the Second World War has been spectacular, with the global catch rising from 20 million tons to 90 million tons between 1950 and 2000. Modern fishing boats are high-technology death machines, which cut through habitats and rake ocean floors to capture valuable commodities. The proportions of by-catch, unintentionally captured fish useless for the market, are huge: 40.4 per cent globally.[7] And so, extinction.

It is, likewise, capitalism that has converted 90 per cent of the earth's wetlands for agriculture[8] and raised extraction to an untenable throughput of 60 billion tons a year. It is capitalist cattle ranching and monoculture plantations in Latin America and South East Asia that destroyed 100 million hectares of tropical forest between 1980 and 2000. It is because of capitalism that 'the average abundance of native species in most major land-based habitats has fallen by at least 20 per cent [...] since 1900[, while] more than 40 per cent of amphibian species, almost 33 per cent of reef-forming corals and more than a third of all marine mammals are threatened'.[9] It is a truism among earth scientists that only indigenously managed ecologies, spared the dynamic of capital accumulation, have avoided the worst.

IV.

Is animal liberation really possible? Are animals, to borrow the animal rights philosopher Ted Benton's terminology, moral agents or moral patients?[10] They can obviously be moral patients insofar as they have sufficient sentience to deserve moral consideration of their desires, not just their presumed needs. It seems clear to me that their oppressive treatment is a political issue. And we are learning more just how much communication is possible with them. But does their intelligence give them the ability to have a say? Could they, for instance, be agents in preventing their own destruction? Could there be an 'interspecies democracy', or a 'zoopolis'?

What is the political meaning of differences in evolved intelligence? We may not be in a good position to answer that question, because, as de Waal implies, we're probably not yet smart enough to know how smart animals are. We have been involved in complex social relations with animals for millennia, sometimes reciprocal, generally exploitative, while having little idea of how large their emotional and communicative world can be.

Nonetheless, to conceptualize these differences, and to replace the *scala naturae*, de Waal gives us the image of a bush. Not one final destination, but several dissimilar paths of development. On one branch, humans; on another, octopuses. This would suggest that the differences between animals are more incremental, distinctions of degree and shade, than metaphysical. And so Montaigne thought: the difference between two humans is greater than that between a human and a beast. And Darwin, an intensely sympathetic observer of animals, agreed. On this basis, he looked forward in *The Descent of Man* to the widening of 'sympathy beyond the confines of man' as a logical moral progression from human universalism.

However, Darwin's argument still implies a human custodial relationship with other animals. As does the example of the golden lion tamarins, and all the other endangered species being conserved and reintroduced into the wild. We might think of it in Karl Marx's terms: are humans the 'universal' species? Marx was evidently wrong to think that other animals act only under the 'dominion of immediate physical need' and metabolize nature only in relation to immediate peers.[11] However, the mere fact of mass extinction seems like a terrible proof of his suggestion that humans are uniquely able not only to take a species-wide view in their acts of production, but to take the whole of nature as their object. The liberation of other animals could hardly be liberation from the stewardship of human beings.

Indeed, the problem is that humans have not yet assumed a relationship of responsibility to the planet and other species. The capitalist mode of production is simply not the kind of machine that enables our species to be responsible. It necessarily takes the whole of nature as its object, but necessarily takes no responsibility for its effects beyond the extraction of value. So that to take responsibility would mean liberating all species from capitalism.

A NOTE ON
TIPPING POINTS

11 JUNE 2021

There have been numerous articles in recent years suggesting that the threshold for a 'cascade of climate tipping points' is either rapidly approaching, or has already been reached.[1]

In the Arctic, it appears likely that the crossing of some tipping points is inevitable, if indeed some thresholds haven't already been breached. An important 2019 study suggests that some global tipping points, which were once thought to be improbable at temperatures below five degrees above pre-industrial temperatures, may have already been crossed at one degree above pre-industrial temperatures.[2] It also seems likely that there exists a threshold of biodiversity loss beyond which other earth systems which are dependent on biodiversity would be thrown into chaos.[3] This means that, in addition to slow, incremental and deadly ecological change, short of radical action we should expect some drastic and obvious transitions with catastrophic effects.

All scientific idiom is constituted by metaphor, working with the logic of language and the unconscious.[4] The metaphor of the climate tipping point has been used by scientists because it had immediate cut-through with the media once it was introduced in around 2005.[5] It had cut through because the term 'tipping point' was already in widespread linguistic use, and resonated with everyday experience of nonlinear change. We've all seen objects tip over once the centre of gravity passes the balance point. The concept of a climate tipping point generally implies an irreversible nonlinear transformation like the sudden loss of the West Antarctic Ice Sheet. The subsidiary metaphor of the domino-like 'cascade' of tipping points was introduced by Timothy Lenton, one of the earliest adopters of 'tipping point' language.[6] It refers to the interaction of several critical thresholds, such as Amazon deforestation and the resulting droughts, the slowdown of the Atlantic Meridional Overturning Circulation that regulates global temperatures, and the thawing of Siberian permafrost, leading to pulses of methane emissions. One thing after another.

The neglect of 'tipping points' is arguably one of the ways in which the IPCC and world governments have consistently underestimated the dangers of climate change.[7] Whence this intellectual caution? The mechanisms of tipping points are not obscure. The fact that complex systems include thresholds in which even small, invisible changes in initial conditions can produce radical, nonlinear effects was discerned in the nineteenth century by the mathematician Henri Poincaré. The insight has long been incorporated into sociology and the study of complex systems, most famously in the form of the 'butterfly effect', and it has been present in ecological thinking since the 1970s. In part, the conservatism of the IPCC is inevitable, given that the governments which it advises can and do place it under considerable pressure. This puts an onus on playing safe with estimates, and refraining from lines of analysis that may be logical (for example, that the methane released by melting permafrost will probably trigger a positive feedback cycle) but can't be empirically verified.

However, there are also some legitimate problems with the 'tipping point' metaphor. It works in this case by condensing the idea of ecological disaster in an image of sudden, spectacular breakdown. The risk is that this deflects attention from more insidious, incremental processes, to which we adapt as, according to urban legend, a frog does to a pan of slowly heating water. Another risk is that implying irreversibility in all cases is misleading and can lead to a kind of fatalism. It is arguably possible to momentarily breach a threshold and return to stability if there is concerted human action. This would particularly be the case with slow-onset tipping points, which unfold over several centuries, and which also happen to be ones we are almost certain to breach.[8] And yet, whatever you call it, whatever metaphor one uses, radical and nonlinear ecosystem change is just a reality with disastrous consequences.

One more serious difficulty with the idiom of 'tipping points' is that it implies the very widespread vernacular of 'balance'. The language of 'balance' is used, for example, in most religious decrees on climate change. And resonates with a conservative critique of human hubris. From this point of view, the aim is to limit human impact on the planet, on the basis of an induction from past experience that we only make things worse. This is not entirely unreasonable, and there is a case for the precautionary principle in how we organize our metabolic relations with the planet. In this light, the current designs of most geoengineering projects look wildly improvident. One ruefully recalls the Soviet scientists who, no doubt unimpressed by the Siberian cold, thought global warming could be a deliberate, utopian enterprise. Or think of those carp imported to US rivers, as Elizabeth Kolbert reports in *Under a White Sky*, as a form of pest control. Attempting to avoid the use of lethal chemicals, they instead introduced fish species that took over and destroyed ecosystems.

However, aside from this being a joyless, partial and limiting perspective, 'balance' often goes beyond conservation and takes an unpleasantly eco-Malthusian hue, with regard to population control. To really exist in 'balance' with the earth, which regularly chokes off life for the majority of its living organisms, would entail a cruel, brutal kind of existence: an existence one would only ever recommend to other, poorer people. Hence the whole effortful endeavour of human science and technology. And if humanity in its capitalist phase is catalysing such disastrous change, we surely have a responsibility to think through ways in which to catalyse better change, not just to conserve a desirable balance. Let us call it ecophilic or biophilic change: a kind of ecological Eros, whose purpose is the enlargement and enrichment of life. In his 2020 essay for the

Royal Society, 'Tipping Positive Change', Timothy Lenton attempts to use the logic of the tipping point in our favour, to suggests ways in which, with adequate information and warning signs, we can deliberately 'tip back' systems that are going wrong. For example, it has been shown that practices of deliberate tipping can, in some circumstances, be deployed in bleached coral reefs.

What would it be like, though, if we went further than that? What if the point was utopian? What if the point was to use the logic of tipping points to maximize the Edenic potential of the world? With all due precautions, with all due care and caveats, with no deference to industry or respect for capital accumulation, what would be wrong with that?

NOTES

INTRODUCTION: A WORLD GROWN OLD

1 Gerardo Ceballos, Paul R. Ehrlich and Rodolfo Dirzo, 'Population Losses and the Sixth Mass Extinction', *PNAS*, 114/30 (2017); Damien Carrington, 'Plummeting Insect Numbers "Threaten Collapse of nature"', *Guardian*, 10 February 2019; FAO, ITPS, GSBI, SCBD and EC, *State of Knowledge of Soil Biodiversity – Status, Challenges and Potentialities: Summary for Policymakers* (Rome: FAO, 2020); Eduardo E. Zattara and Marcelo A. Aizen, 'Worldwide Occurrence Records Suggest a Global Decline in Bee Species Richness', *One Earth*, 4/1 (2021).

2 Shannon Hall, 'Exxon Knew About Climate Change Almost 40 Years Ago', *Scientific American*, 26 October 2015.

3 Paul Griffin, *CDP Carbon Majors Report 2017*, CDP, July 2017.

4 Jason W. Moore, ed., *Anthropocene or Capitalocene? Nature, History, and the Crisis of Capitalism* (Oakland, CA: PM Press, 2016).

5 Renée Lertzman, *Environmental Melancholia: Psychoanalytic Dimensions of Engagement* (London: Routledge, 2015); see also Sally Weintrobe, ed., *Engaging with Climate Change: Psychoanalytic and Interdisciplinary Perspectives* (London: Routledge, 2012) and Sally Weintrobe, *Psychological Roots of the Climate Crisis: Neoliberal Exceptionalism and the Culture of Uncare* (London: Bloomsbury, 2021); and Anouchka Grose, *A Guide to Eco-Anxiety: How to Protect the Planet and Your Mental Health* (London: Watkins, 2020).

6 Glenn Scherer, 'How the IPCC Underestimated Climate Change', *Scientific American*, 6 December 2012.

7 Fred Pearce, 'As Climate Change Worsens, A Cascade of Tipping Points Looms', *Yale Environment 360*, 5 December 2019.

8 Timothy M. Lenton et al., 'Climate Tipping Points – Too Risky to Bet Against', *Nature* (27 November 2019).

9 David Wallace-Wells, *The Uninhabitable Earth* (London: Allen Lane, 2019).

10 Sarah Emerson, 'What the Paris Climate Agreement Means For Big Oil', *Motherboard*, 4 November 2016.

11 Kelly Levin and Taryn Fransen, 'INSIDER: Why Are INDC Studies Reaching Different Temperature Estimates?', World Resources Institute, 9 November 2015.

12 George Orwell, 'Why I Write', in *Facing Unpleasant Facts: Narrative Essays* (Boston, MA: Mariner Books, 2009).

13 See Niclas Svenningsen, 'Aviation, Offsets and the Paris Agreement', *ICAO Environmental Report 2016* and Jocelyn Timperly, 'Corsia: The UN's Plan to "Offset" Growth in Aviation Emissions', Carbon Brief, 4 February 2019.

14 Natalie Sachmechi, 'Airlines Are Being Bailed Out Again, Here's What Economists Think Will Happen Next', *Forbes*, 17 April 2020; Sandra Laville, 'Coronavirus: Airlines Seek €12.8bn in Bailouts Without Environmental Conditions Attached', *Guardian*, 27 April 2020.

15 See David Robinson Simon, *Meatonomics: How the Rigged Economics of Meat and Dairy Make You Consume Too Much And How to Eat Better, Live Longer, and Spend Smarter* (Newburyport, MA: Conari Press, 2013); Vaclav Smil, *Should We Eat Meat?: Evolution and Consequences of Modern Carnivory* (London: John Wiley & Sons, 2013), pp. 170 and 227–8.

16 Zhang Chun, 'China's New Blueprint for an "Ecological Civilization"', *Diplomat*, 30 September 2015.

17 Richard Smith, *China's Engine of Environmental Collapse* (London: Pluto Press, 2020).

LETTING GO

1 John Berger, 'Undefeated Despair', *Critical Inquiry*, 32/4 (2006).

DROPPING LIKE BEES

1 'Pesticides Could Wipe Out Bumblebee Populations, Study Shows', *Guardian*, 14 August 2017.

2 Theodor W. Adorno and Max Horkheimer, *Dialectic of Enlightenment* (London: Verso, 2016); Carolyn Merchant, *The Death of Nature: Women, Ecology, and the Scientific Revolution* (San Francisco: HarperSanFrancisco, 1990).

3 *Capitalism in the Web of Life: Ecology and the Accumulation of Capital* (London: Verso, 2015).

4 'The Capitalocene, Part II: Abstract Social Nature and the Limits to Capital', https://www.researchgate.net/publication/264457281_The_Capitalocene_Part_II_Abstract_Social_Nature_and_the_Limits_to_Capital.

5 Jason W. Moore, 'The Rise of Cheap Nature', in *Anthropocene or Capitalocene?*

OPARIN'S OCEAN

1 Charles Lyell, *Principles of Geology* (London: Penguin, 2005).

2 Bernard Kettlewell, *The Evolution of Melanism: The Study of a Recurring Necessity, with Special Reference to Industrial Melanism in the Lepidoptera* (Oxford: OUP, 1973).

3 Yasemin Saplakoglu, 'Humans May Be Influencing Bird Evolution in Their Backyards', *Scientific American*, 19 October 2017.

4 Letter of 1 February 1871[?], www.darwinproject.ac.uk.

5 A.I. Oparin, *The Origin of Life* (Moscow: Foreign Languages Publishing House, 1955).

UNWORLDLINESS

1 Nietzsche, *Ecce Homo*, Penguin, 2004, p. 101.

2 Quoted in Enzo Traverso, *Left-Wing Melancholia: Marxism, History and Memory* (London: Verso, 2017), p. 225.

3 Extract from the *Future of an Illusion* in Peter Gay, ed., *The Freud Reader* (New York: W.W. Norton & Co., 1995), p. 693.

4 Freud, *Civilization and Its Discontents* (London: Penguin, 2002), p. 105.

5 Donald W. Winnicott, *Playing and Reality* (London: Tavistock Publications, 1971); Michael Eigen, *Toxic Nourishment* (London: Karnac Books, 1999).

6 Jason W. Moore in *Anthropocene or Capitalocene?*

7 Carl Safina, 'Thinking in the Deep: Inside the Mind of an Octopus', *New York Times*, 27 December 2016.

8 Dana Gioia, 'Poetry as Enchantment', *Dark Horse*, 34 (2015).

9 'Dream Kitsch: Gloss on Surrealism', in *The Work of Art in the Age of Its Technological Reproducibility, and Other Writings on Media*, ed. Michael W. Jennings, Brigid Doherty and Thomas Y. Levin (Cambridge, MA: Harvard University Press, 2008).

10 'Why Does the Russian Revolution Matter?', 6 May 2017.

THE ATOMIC GENIE

1 Keith Barnham, *The Burning Answer: A User's Guide to the Solar Revolution* (London: Weidenfeld & Nicolson, 2014).

2 'More Solar Power Hurts Nuclear Energy. But It Also Hurts Itself', *Economist*, 8 September 2018.

3 Suzanne Waldman, 'Timeline: The IPCC's Shifting Position on Nuclear Energy', *Bulletin of the Atomic Scientists*, 8 February 2015; Aviel Verbruggen and Erik Laes, 'Sustainability Assessment of Nuclear Power: Discourse Analysis of IAEA and IPCC Frameworks', *Environmental Science & Policy*, 51 (2015), pp. 170–180; IAEA, *Methodology for the Assessment of Innovative Nuclear Reactors and Fuel Cycles Report of Phase 1B (first part) of the International Project on Innovative Nuclear Reactors and Fuel Cycles (INPRO)*, IAEA-TECDOC-1434 (Vienna: IAEA, 2004).

4 Committee on Climate Change, *The Renewable Energy Review* (London: CCC, 2011).

5 Andy Stirling and Phil Johnstone, 'Why Is the UK Government So Infatuated With Nuclear Power?', *Guardian*, 29 March 2018.

6 International Institute for Sustainable Development, 'The United Kingdom Is to Subsidize Nuclear Power—But At What Cost?', https://www.iisd.org/story/the-united-kingdom-is-to-subsidize-nuclear-power-but-at-what-cost/.

7 Monbiot, 'Why Must UK Have to Choose Between Nuclear and Renewable Energy?', *Guardian*, 27 May 2011; Phillips, 'People's Fission', *New Republic*, 14 April 2016.

8 'Nuclear Power – Yes Please. Hinkley Point – No Thanks', 15 September 2016.

9 Michael Dittmar, 'The End of Cheap Uranium', World Resources Forum lecture, Davos, 20 September 2011.

10 Antoine Monnet, Sophie Gabriel and Jacques Percebois, 'Long-Term Availability of Global Uranium Resources', *Resources Policy*, 53 (2017).

11 James Conca, 'Uranium Seawater Extraction Makes Nuclear Power Completely Renewable', *Forbes*, 1 July 2016.

12 Monbiot, 'The Nuclear Industry Stinks. But That Is Not a Reason to Ditch Nuclear Power', *Guardian*, 4 July 2011.

13 Teri Sforza, 'Watchdog: Nuclear Waste Can Be Stored at New San Onofre Site, Coastal Commission Says', *Orange County Register*, 7 October 2015.

14 See United States Nuclear Regulatory Commission, 'San Onofre Nuclear Generating Station, Units 1, 2, and 3 and the Independent Spent Fuel Storage Installation – Issuance of Amendments re: Changes to the Emergency Plan', 5 June 2015, https://publicwatchdogs.org/wp-content/uploads/2016/08/SONGS-ISFSI-NRC-Wengert-to-Palmisano-SCE-6-5-15-1.pdf and Nina Babiarz, 'Public Watchdogs Exposes Shocking Emergency Response Plans at SONGS', Public Watchdogs, 30 September 2016.

15 Mitch Jacoby, 'As Nuclear Waste Piles Up, Scientists Seek the Best Long-Term Storage Solutions', *Chemical & Engineering News*, 30 March 2020.

16 James Temperton, 'Inside Sellafield: How the UK's Most Dangerous Nuclear Site Is Cleaning Up Its Act', *Wired*, 17 September 2016 and Ruth Quinn, 'Sellafield "riddled with safety flaws", according to BBC investigation', *Guardian*, 5 September 2016.

17 Vincent Lalenti, *Deep Time Reckoning: How Future Thinking Can Help Earth Now* (Cambridge, MA: MIT Press, 2020).

18 See Paul Brown, 'First Nuclear Power Plant to Close', *Guardian*, 21 March 2003 and 'Cheap and Lethal Nuclear By-Product', ibid., 12 January 2001.

19 'UK Labour Party Split Over Nuclear Power', *Financial Times*, 6 May 2018.

20 Geoff Mann and Joel Wainwright, *Climate Leviathan: A Political Theory of Our Planetary Future* (London: Verso, 2017).

21 Monbiot, 'Why Must UK Have to Choose Between Nuclear and Renewable Energy?', 27 May 2011; Porritt, 'Why the UK Must Choose Renewables Over Nuclear: An Answer to Monbiot', 26 July 2011.

22 S. Schlömer, 'Annex III: Technology-Specific Cost and Performance Parameters', in *Climate Change 2014: Mitigation of Climate Change. Contribution of Working Group III to the Fifth Assessment Report of the Intergovernmental Panel on Climate Change*, ed. O. Edenhofer et al. (Cambridge: CUP, 2014).

23 Manfred Lenzen, 'Life Cycle Energy and Greenhouse Gas Emissions of Nuclear Energy: A Review', *Energy Conversion and Management*, 49 (2008) and Ethan S. Warner and Garvin A. Heath, 'Life Cycle Greenhouse Gas Emissions of Nuclear Electricity Generation', *Journal of Industrial Ecology*, 16 (2012).

24 Benjamin K. Sovacool, 'Valuing the Greenhouse Gas Emissions from Nuclear Power: A Critical Survey', *Energy Policy*, 36 (2008).

25 'Energy Balance of Nuclear Power Generation: Life Cycle Analysis of Nuclear Power – A Summary', https://www.energyagency.at/fileadmin/dam/pdf/publikationen/berichteBroschueren/Endbericht_LCA_Nuklearindustrie-engl.pdf.

26 Monbiot, 'Why Fukushima Made Me Stop Worrying and Love Nuclear Power', *Guardian*, 21 March 2011, and 'Why must UK have to choose between nuclear and renewable energy?', *Guardian*, 27 May 2011.

WHAT, OR WHOM, WILL WE EAT?

1 Damian Carrington, 'True Cost of Cheap Food Is Health and Climate Crises, Says Commission', *Guardian*, 16 July 2019.

2 'Two Billion People Without Access to Healthy Food: UN', *Al Jazeera*, 15 July 2019 and Sarah Boseley, 'World Hunger on the Rise As 820m at Risk, UN Report Finds', *Guardian*, 15 July 2019.

3 Heinz-Wilhelm Strubenhoff, 'Can 10 Billion People Live and Eat Well on the Planet? Yes.', Brookings, 28 April 2015, https://www.brookings.edu/blog/future-development/2015/04/28/can-10-billion-people-live-and-eat-well-on-the-planet-yes/.

4 Raj Patel and Jason W. Moore, *A History of the World in Seven Cheap Things: A Guide to Capitalism, Nature, and the Future of the Planet* (Berkeley, CA: University of California Press, 2018).

5 Jonathan Watts, 'Amazon Deforestation Accelerating towards Unrecoverable "Tipping Point"', *Guardian*, 25 July 2019.

6 Damian Carrington, 'True Cost of Cheap Food'.

7 Jason W. Moore, 'Cheap Food and Bad Climate: From Surplus Value to Negative Value in the Capitalist World-Ecology', *Critical Historical Studies*, 2/1 (2015).

8 Bill Gates, 'Food Is Ripe for Innovation', Mashable, 21 March 2013; S. Jagtap and S. Rahimifard, 'The Digitisation of Food Manufacturing to Reduce Waste – Case Study of a Ready Meal Factory', *Waste Management*,

87 (2019); and Andrew Balmford et al., 'The Environmental Costs and Benefits of High-Yield Farming', *Nature Sustainability*, 1 (2018).

9 Troy Vettese, 'To Freeze the Thames: Natural Geo-Engineering and Biodiversity', *New Left Review*, 111 (2018).

A NOTE ON CLIMATE SADISM

1 Becca Warner, 'The End of the Amazon', *Ecologist*, 15 October 2018, https://theecologist.org/2018/oct/15/end-amazon.

2 Sam Cowie, 'Jair Bolsonaro Praised the Genocide of Indigenous People. Now He's Emboldening Attackers of Brazil's Amazonian Communities', *Intercept*, 16 Feb 2019.

3 Sunrise Movement, Twitter, 21 August 2019.

4 Jonathan Watts, 'Jair Bolsonaro Claims NGOs Behind Amazon Forest Fire Surge – But Provides No Evidence', *Guardian*, 21 August 2019.

THE SUN NOW EMBRACES NATURE

1 See Stephen J. Pyne, *Fire in America: A Cultural History of Wildland and Rural Fire* (Seattle, WA: University of Washington Press, 1982); *Burning Bush: A Fire History of Australia* (ibid., 1991); *Vestal Fire: An Environmental History, Told Through Fire, of Europe and Europe's Encounter with the World* (ibid., 1997); *World Fire: The Culture of Fire on Earth* (ibid., 1995); and *The Pyrocene: How We Created an Age of Fire, and What Happens Next* (Berkeley, MA: University of California Press, 2021).

2 Mike Davis, *Ecology of Fear* (New York: Metropolitan Books, 1998).

3 Alfred W. Crosby, *Ecological Imperialism: The Biological Expansion of Europe, 900–1900* (Cambridge: CUP, 1986).

4 'Australia Fires: A Visual Guide to the Bushfire Crisis', BBC News, 31 January 2020.

5 See e.g. W. Matt Jolly et al., 'Climate-Induced Variations in Global Wildfire Danger from 1979 to 2013', *Nature Communications*, 6 (2015).

6 Kaitlin M. Keegan et al., 'Climate Change and Forest Fires Synergistically Drive Widespread Melt Events of the Greenland Ice Sheet', *PNAS*, 111/22 (2014).

7 'Arctic Wildfires: How Bad Are They and What Caused Them?', BBC News, 2 August 2019.

8 Danilo Mollicone, Hugh D. Eva and Frédéric Achard, 'Human Role in Russian Wildfires', *Nature* 440 (2006), pp. 436–7.

9 Stephen Pyne, 'California Wildfires Signal the Arrival of a Planetary Fire Age', *Conversation*, 1 November 2019.

THE DARK SIDE OF CARBON DEMOCRACY

1 Greta Thunberg, 'Our House Is Still on Fire and You're Fuelling the Flames', World Economic Forum lecture, 21 January 2020.

2 David Shearman and Joseph Wayne Smith, *The Climate Change Challenge and the Failure of Democracy* (Westport, CT: Praeger, 2007).

3 Nathaniel Rich, 'Losing Earth: The Decade We Almost Stopped Climate Change', *New York Times*, 1 August 2018.

4 Simon Kuper, 'The Myth of Green Growth', *Guardian*, 24 October 2019.

5 Alyssa Battistoni and Jedediah Britton-Purdy, 'After Carbon Democracy', *Dissent*, Winter 2020.

6 Timothy Mitchell, *Carbon Democracy: Political Power in the Age of Oil* (London: Verso, 2011).

7 Michael Mann, *The Dark Side of Democracy: Explaining Ethnic Cleansing* (Cambridge: CUP, 2005).

8 Rosa Luxemburg, *The Crisis of Social Democracy, with an Appendix: Guidelines for the Tasks of International Social Democracy* (Zurich: Union, 1916).

WHAT IS AN IDEOLOGY WITHOUT A SPACE?

1 Philip Oltermann, 'German Far Right Infiltrates Green Groups with Call to Protect the Land', *Guardian*, 28 June 2020; Sarah Manavis, 'Coronavirus Crisis Gives Eco-Fascism a Boost', *Financial Review*, 12 June 2020; and Beth Gardiner, 'White Supremacy Goes Green', *New York Times*, 28 February 2020.

2 Oltermann, 'German Far Right Infiltrates Green Groups'.

3 Beth Gardiner, 'White Supremacy Goes Green'.

4 Quoted in Blair Taylor, 'Ecofascism and Far-Right Environmentalism in the United States', in Bernhard Forchtner, ed., *The Far Right and the Environment: Politics, Discourse and Communication* (London: Routledge, 2020).

5 'Le Pen's National Rally Goes Green in Bid for European Election Votes', *France24*, 20 April 2019.

6 See Julian Göpffarth, 'Why Did Heidegger Emerge as the Central Philosopher of the Far Right?', openDemocracy, 23 June 2020; Matthew Connelly, *Fatal Misconception: The Struggle to Control World Population* (Cambridge, MA: Harvard University Press, 2008); Janet Biehl and Peter Staudenmaier, *Ecofascism Revisited* (Porsgrunn, Norway: New Compass, 2011); and Jedediah Purdy, 'Environmentalism's Racist History', *New Yorker*, News, 13 August 2015.

7 Daniel Jones, 'Greenshirts – The (Mis)use of Environmentalism by the Extreme Right', History Workshop, 21 April 2020, https://www.historyworkshop.org.uk/greenshirts-the-misuse-of-environmentalism-by-the-extreme-right/.

8 See Paul Kingsnorth, 'The Lie of the Land: Does Environmentalism Have a Future in the Age of Trump?', *Guardian*, 18 March 2017, and 'England's Uncertain Future', *Guardian*, 13 March 2015.

9 Damien Gayle, 'Does Extinction Rebellion Have a Race Problem?', *Guardian*, 4 October 2019.

10 Andreas Malm and The Zetkin Collective, *White Skin, Black Fuel: On the Danger of Fossil Fascism* (London: Verso, 2021).

11 Quoted in Ian Angus and Simon Butler, *Too Many People? Population, Immigration, and the Environmental Crisis* (Chicago: Haymarket, 2011); see also his recapitulation of Paul Ehrlich's eco-Malthusianism in Dave Foreman and Laura Carroll, *Man Swarm: How Overpopulation Is Killing the Wild World* (LiveTrue, 2014).

ULTIMA THULE: AN OBITUARY

1 In Alexis S. Troubetzkoy, *Arctic Obsession: The Lure of the Far North* (New York: Thomas Dunne Books, 2011).

2 Elisha Kent Kane, *The U.S. Grinnell Expedition in Search of Sir John Franklin: A Personal Narrative* (Cambridge: CUP, 2015), p. 127.

3 Many of the facts and references in this passage come from Stephen J. Pyne's spectacular book *The Ice: A Journey to Antarctica* (Seattle, WA: University of Washington Press, 1986).

4 Andrew E. Derocher, *Polar Bears: A Complete Guide to Their Biology and Behaviour* (Baltimore, MD: Johns Hopkins University Press, 2012).

5 *The Alaska Account of John Muir: Travels in Alaska, The Cruise of the Corwin, Stickeen & Alaska Days with John Muir* (Prague: e-artnow, 2015), p. 436.

6 Oliver Milman, 'Greenhouse Gas Emissions Transforming the Arctic into "an Entirely Different Climate"', *Guardian*, 8 Dec 2020.

7 Mark Serreze, quoted in Peter Wadhams, *A Farewell to Ice: A Report from the Arctic* (London: Allen Lane, 2017), p. 4.

8 Ossie Michelin, '"Solastalgia": Arctic Inhabitants Overwhelmed by New Form of Climate Grief', *Guardian*, 15 October 2020 and Livia Albeck-Ripka, 'Why Lost Ice Means Lost Hope for an Inuit Village', *New York Times*, 25 November 2017.

9 John Keats, Letter to George and Georgiana Keats, Hampstead, 18[?] December 1818, available at www.keats-poems.com.

10 Apsley Cherry-Garrard, *The Worst Journey in the World* (London: Vintage Classics, 2010); Lisle A. Rose, *Explorer: The Life of Richard E. Byrd* (Columbia, MO: University of Missouri Press, 2013).

DISASTER AND DENIALISM

1 See E.J. Dickson, 'How the Right Spread a False Rumour About Antifa and Wildfires', *Rolling Stone*, 11 September 2020, and Andreas Malm and The Zetkin Collective, *White Skin, Black Fuel*.

2 Oliver Milman, 'Heating Arctic May Be to Blame for Snowstorms in Texas, Scientists Argue', *Guardian*, 17 February 2021 and 'Fact Check: The Causes for Texas' Blackout Go Well Beyond Wind Turbines', *Reuters*, 19 February 2021.

3 Jesse Jenkins, Twitter, 15 February 2021.

4 Erin Douglas and Ross Ramsey, 'No, Frozen Wind Turbines Aren't the Main Culprit for Texas' Power Outages', *Texas Tribune*, 16 February 2021; Marcy de Luna and Amanda Drane, 'What Went Wrong with the Texas Power Grid?', *Houston Chronicle*, 16 February 2021; 'Texas Weather: Deaths Mount as Winter Storm Leaves Millions without Power', BBC News, 17 February 2021; and Alex Samuels, 'Nearly 12 Million Texans Now Face Water Disruptions. The State Needs Residents to Stop Dripping Taps', *Texas Tribune*, 17 February 2021.

5 J. Cohen et al., 'Divergent Consensuses on Arctic Amplification Influence on Midlatitude Severe Winter Weather', *Nature Climate Change*, 10 (2020).

6 Asher Price and Bob Sechler, 'Did Texas Energy Regulators Fail to Mandate Winter Protections?', *Austin American-Statesman*, 17 February 2021 and Ari Natter and Jennifer A. Dlouhy, 'Texas Was Warned a Decade Ago Its Grid Was Unready for Cold', Bloomberg, 17 February 2021.

7 'State of Texas Emergency Management Plan: Basic Plan', February 2020, see www.tdem.texas.gov.

8 Alexandra Villarreal and Erum Salam, 'The Texans Facing Blackouts and Burst Pipes: "Do I Wait for the Ceiling to Cave In?"', *Guardian*, 18 February 2021.

9 See https://www.whitehouse.gov/briefing-room/statements-releases/2021/02/14/president-joseph-r-biden-jr-approves-texas-emergency-declaration/.

10 See Katie Shepherd, 'Rick Perry Says Texans Would Accept Even Longer Power Outages "to Keep the Federal Government out of Their Business"', *Washington Post*, 18 February 2021.

11 Ben Piven, 'In Extreme Texas Cold, Green New Deal Turns into Hot Potato', *Al Jazeera*, 20 February 2021; see 'Chris Hayes Debunks GOP, Right-Wing Media Lies about Texas Power Outages', MSNBC, 17 February 2021.

12 'What's Up in Texas?', https://www.republicanleader.gov/whats-up-in-texas/.

13 Lois Beckett, 'Older People Would Rather Die Than Let Covid-19 Harm US Economy – Texas Official', *Guardian*, 24 March 2020.

14 Christopher Brito, 'Texas Mayor Resigns After Telling Residents Desperate for Power and Heat "Only the Strong Will Survive"', CBS, 18 February 2021.

15 'Petro-Masculinity: Fossil Fuels and Authoritarian Desire', *Millennium*, 47/1 (2018), pp. 25–44.

NOTHING BESIDE REMAINS

1 Christopher J. Bickerton and Carlo Invernizzi Accetti, *Technopopulism: The New Logic of Democratic Politics* (Oxford: OUP, 2021).

2 Erica Chenoweth and Maria J. Stephan, *Why Civil Resistance Works: The Strategic Logic of Nonviolent Conflict* (New York: Columbia University Press, 2011).

3 Vikram Dodd and Jamie Grierson, 'Terrorism Police List Extinction Rebellion as Extremist Ideology', *Guardian*, 10 January 2020.

4 'Major Shift in UK Land Use Needed to Deliver Net Zero Emissions', CCC, 23 January 2020, https://www.theccc.org.uk/2020/01/23/major-shift-in-uk-land-use-needed-to-deliver-net-zero-emissions/.

5 'Longannet Carbon Capture Scheme Scrapped', BBC News, 19 October 2011; Tim Webb, 'E.ON Shelves Plans to Build Kingsnorth Coal Plant', *Guardian*, 20 October 2010.

6 Julian M. Allwood et al., *Absolute Zero*, 2019, https://www.repository.cam.ac.uk/handle/1810/299414.

7 On the necessity and dangers of CCS, see Andreas Malm and Wim Carton, 'Seize the Means of Carbon Removal: The Political Economy of Direct Air Capture', *Historical Materialism*, 29/1 (2021).

8 Jean-Paul Deléage, Jean-Claude Debeir and Daniel Hémery, *In the Servitude of Power: Energy and Civilization Through the Ages*, Zed Books, 1991.

9 'The Work/Energy Crisis in the Apocalypse', in George Caffentzis, *In Letters of Blood and Fire: Work, Machines, and Value in the Bad Infinity of Capitalism* (Oakland, CA: PM Press, 2013).

10 Matthew T. Huber, 'Energizing Historical Materialism: Fossil Fuels, Space and the Capitalist Mode of Production', *Geoforum*, 40/1 (2009).

11 Bent Sørensen, *Renewable Energy: Physics, Engineering, Environmental Impacts, Economics and Planning*, 4th edn (Oxford: Academic Press, 2010).

12 Bruce Usher, *Renewable Energy: A Primer for the Twenty-First Century* (New York: Columbia University Press, 2019).

13 'Oregon Moving to Center of Wave Energy Development', Oregon State University Newsroom, 18 June 2009.

WHAT THE EYE LIKES

1 See Mark Kaufman, 'The Carbon Footprint Sham', Mashable, 13 July 2020.

2 Kathryn Tanner, *Christianity and the New Spirit of Capitalism* (New Haven, CT: Yale University Press, 2019).

3 Lisa Holland, 'Climate Change: Revealed – How Many Britons Are Unwilling to Change Their Habits to Tackle the Crisis', Sky News, 7 April 2021.

4 Vikram Dodd and Jamie Grierson, 'Terrorism Police List Extinction Rebellion as Extremist Ideology'.

5 Dave Fawbert, '"Eco-Anxiety": How to Spot It and What to Do About It', BBC News, 27 March 2019.

6 Catherine Keller, *Political Theology of the Earth: Our Planetary Emergency and the Struggle for a New Public* (New York: Columbia University Press, 2018).

7 Denis Campbell, 'UK Has Experienced "Explosion" in Anxiety Since 2008, Study Finds', *Guardian*, 14 September 2020.

8 Christopher Bollas, *The Shadow of the Object: Psychoanalysis of the Unthought Known* (London: Routledge, 2018).

9 Marion Milner, *On Not Being Able to Paint* (London: Routledge, 2010).

SCALA NATURAE

1 Matthew Chrulew, 'Saving the Golden Lion Tamarin', in Deborah Bird Rose et al., eds, *Extinction Studies Stories of Time, Death, and Generations* (New York: Columbia University Press, 2017).

2 Eva Meijer, *Animal Languages: The Secret Conversations of the Living World* (London: John Murray Press, 2019) and *When Animals Speak: Toward an Interspecies Democracy* (New York: New York University Press, 2019); Carl Safina, *Becoming Wild: How Animals Learn to Be Animals* (London: Oneworld Publications, 2020) and *Beyond Words: What Animals Think and Feel* (New York: Macmillan, 2015); Peter Godfrey-Smith, *Other Minds: The Octopus, the Sea, and the Deep Origins of Consciousness* (New York: Farrar, Straus & Giroux, 2016); Peter Wohlleben, *The Inner Life of Animals: Surprising Observations of a Hidden World* (London: Vintage, 2018); Frans de Waal, *Are We Smart Enough to Know How Smart Animals Are?* (New York: W. W. Norton & Co, 2016); Marc Bekoff, *The Emotional Lives of Animals: A Leading Scientist Explores Animal Joy, Sorrow, and Empathy – and Why They Matter* (Novato, CA: New World Library, 2008); and Donald R. Griffin, *Animal Minds: Beyond Cognition to Consciousness* (Chicago: University of Chicago Press, 2001).

3 Martin Heidegger, *The Fundamental Concepts of Metaphysics: World, Finitude, Solitude*, trans. W. McNeill and N. Walker (Bloomington, IN: Indiana University Press, 1995), p. 196.

4 *Salvage #7: Tragedy of the Worker*, October 2019.

5 George Monbiot, 'The RSPCA Rescues One Seal – and Condones the Killing of Many Others', *Guardian*, 19 September 2018.

6 Ashley Dawson, *Extinction: A Radical History* (New York: OR Books, 2016).

7 R.W.D. Davies et al., 'Defining and Estimating Global Marine Fisheries Bycatch', *Marine Policy* 33/4 (2009), pp. 661–72.

8 Sir Robert Watson, in *Extinction: The Facts*, BBC One, 13 September 2020.

9 IPBES, 'Nature's Dangerous Decline "Unprecedented"; Species Extinction Rates "Accelerating"', https://ipbes.net/news/Media-Release-Global-Assessment.

10 Ted Benton, *Natural Relations: Ecology, Animal Rights and Social Justice* (London: Verso, 1993).

11 'Estranged Labour', *Economic and Philosophical Manuscripts of 1844*, see https://www.marxists.org/archive/marx/works/1844/manuscripts/labour.htm.

A NOTE ON TIPPING POINTS

1 See Damian Carrington, 'Climate Emergency: World "May Have Crossed Tipping Points"', *Guardian*, 27 November 2019 and Will Steffen et al., 'Trajectories of the Earth System in the Anthropocene', *PNAS*, 115/33 (2018).

2 Timothy M. Lenton et al., 'Climate Tipping Points'.

3 Jeremy Hance, 'Could biodiversity destruction lead to a global tipping point?', *Guardian*, 16 January 2018.

4 Giles Foden, 'Skittles: The Story of the Tipping Point Metaphor and Its Relation to New Realities', in Timothy O'Riordan and Timothy Lenton, *Addressing Tipping Points for a Precarious Future* (Oxford: OUP, 2013).

5 Sandra van der Hel, 'Tipping Points and Climate Change: Metaphor Between Science and the Media', *Environmental Communication*, 12/5 (2018).

6 Fred Pearce, 'As Climate Change Worsens, A Cascade of Tipping Points Loom', Yale Environment 360, 5 December 2019, https://e360.yale.edu/features/as-climate-changes-worsens-a-cascade-of-tipping-points-looms.

7 Timothy M. Lenton et al., 'Climate Tipping Points'; Donald A. Brown, 'Lessons Learned from IPCC's Underestimation of Climate Change Impacts about the Need for a Precautionary Climate Change

Science', in L. Westra, K. Bosselmann and M. Fermeglia, eds, *Ecological Integrity in Science and Law* (Cham, Switzerland: Springer, 2020); and Glenn Scherer, 'How the IPCC Underestimated Climate Change'.
8 Paul D. L. Ritchie et al., 'Overshooting Tipping Point Thresholds in a Changing Climate', *Nature*, 592 (2021), p. 517.

ACKNOWLEDGEMENTS

These essays were written over a period of five years, so there are more people to thank for their input and influence than I can possibly name.

However, I would like to specifically acknowledge and thank my friends on the Salvage collective — Jamie Allinson, China Miéville and Rosie Warren — for providing sources, reading essays, giving encouragement, and above all for supplying the inimitable collective unconscious without which these essays wouldn't have been produced.

I would also like to thank Susie Nicklin at Indigo for appreciating these essays enough to publish them, and my agent, Karolina Sutton, for helping make this happen. I'm also grateful to my supporters on Patreon.

THE

INDIGO

PRESS

The Indigo Press is an independent publisher of contemporary fiction and non-fiction, based in London. Guided by a spirit of internationalism, feminism and social justice, we publish books to make readers see the world afresh, question their behaviour and beliefs, and imagine a better future.

Browse our books and sign up to our newsletter for special offers and discounts:

theindigopress.com

The Indigo Press app brings our books and exclusive bonus content to readers around the world. Available now on iOS, Android, and your web browser.

Follow *The Indigo Press* on social media for the latest news, events and more:

@PressIndigoThe

@TheIndigoPress

@TheIndigoPress

The Indigo Press

@theindigopress

Transforming a manuscript into the book
you hold in your hands is a group project.

Richard would like to thank everyone who
helped to publish *The Disenchanted Earth*.

THE INDIGO PRESS TEAM

Susie Nicklin
Phoebe Barker
Honor Scott

JACKET DESIGN

Michael Salu
Luke Bird

PUBLICITY

Claire Maxwell

FOREIGN RIGHTS

The Marsh Agency

EDITORIAL PRODUCTION

Tetragon
Gesche Ipsen